T0206184

Springer Aerospace Technology

Series Editors

Sergio De Rosa, DII, University of Naples Federico II, Napoli, Italy

Yao Zheng, School of Aeronautics and Astronautics, Zhejiang University, Hangzhou, Zhejiang, China

Elena Popova, Air Navigation Bridge Russia, Chelyabinsk, Russia

The series explores the technology and the science related to the aircraft and spacecraft including concept, design, assembly, control and maintenance. The topics cover aircraft, missiles, space vehicles, aircraft engines and propulsion units. The volumes of the series present the fundamentals, the applications and the advances in all the fields related to aerospace engineering, including:

- structural analysis,
- aerodynamics,
- aeroelasticity,
- aeroacoustics,
- flight mechanics and dynamics
- orbital maneuvers,
- avionics,
- systems design,
- materials technology,
- launch technology,
- payload and satellite technology,
- space industry, medicine and biology.

The series' scope includes monographs, professional books, advanced textbooks, as well as selected contributions from specialized conferences and workshops.

The volumes of the series are single-blind peer-reviewed.

To submit a proposal or request further information, please contact: Mr. Pierpaolo Riva at pierpaolo.riva@springer.com (Europe and Americas) Mr. Mengchu Huang at mengchu.huang@springer.com (China)

The series is indexed in Scopus and Compendex

Georgy Alekseevich Kryzhanovsky ·
Anatoly Ivanovich Kozlov · Oleg Ivanovich Sauta ·
Yuri Grigoryevich Shatrakov ·
Ivan Nikolaevich Shestakov

Modeling of Transportation Aviation Processes

Springer

Georgy Alekseevich Kryzhanovsky
Saint Petersburg, Russia

Anatoly Ivanovich Kozlov
Moscow, Russia

Oleg Ivanovich Sauta
Saint Petersburg, Russia

Yuri Grigoryevich Shatrakov
Saint Petersburg, Russia

Ivan Nikolaevich Shestakov
Saint Petersburg, Russia

ISSN 1869-1730 ISSN 1869-1749 (electronic)
Springer Aerospace Technology
ISBN 978-981-19-7609-4 ISBN 978-981-19-7607-0 (eBook)
https://doi.org/10.1007/978-981-19-7607-0

© The Editor(s) (if applicable) and The Author(s), under exclusive license to Springer Nature Singapore Pte Ltd. 2023
This work is subject to copyright. All rights are solely and exclusively licensed by the Publisher, whether the whole or part of the material is concerned, specifically the rights of translation, reprinting, reuse of illustrations, recitation, broadcasting, reproduction on microfilms or in any other physical way, and transmission or information storage and retrieval, electronic adaptation, computer software, or by similar or dissimilar methodology now known or hereafter developed.
The use of general descriptive names, registered names, trademarks, service marks, etc. in this publication does not imply, even in the absence of a specific statement, that such names are exempt from the relevant protective laws and regulations and therefore free for general use.
The publisher, the authors, and the editors are safe to assume that the advice and information in this book are believed to be true and accurate at the date of publication. Neither the publisher nor the authors or the editors give a warranty, expressed or implied, with respect to the material contained herein or for any errors or omissions that may have been made. The publisher remains neutral with regard to jurisdictional claims in published maps and institutional affiliations.

This Springer imprint is published by the registered company Springer Nature Singapore Pte Ltd.
The registered company address is: 152 Beach Road, #21-01/04 Gateway East, Singapore 189721, Singapore

Concepts, tasks, methods and types of modeling in the study of transport processes and the assessment of their safety are given. Mathematical models of creation and organization of transport space, transport networks are considered. Special attention is paid to modeling the activities of operators, LPR and collectives. The methods of multi-criteria optimization of decision-making processes are given, taking into account the motivation of operators of transport processes of their LPR.

Introduction

A system is usually understood as an ordered set of objects (elements) united by any connections, the interaction of which is aimed at achieving a certain goal.

A transport system can be called a set of objects of transport space, transport equipment and people united for the implementation of the transport process.

The principal feature of aviation transport systems in comparison with other production systems is that the transport process is carried out within a single transport space. This circumstance determines the need for an orderly unification of individual transport systems into an organization formed in a certain way. The task of studying processes in transport systems should be solved taking into account such characteristic properties and features as integrity, hierarchy, purposefulness (activity), the presence of uncertainty in the conditions of complex dynamic transport situations, competition and rather strict time limits on all decision-making processes.

The listed features of aviation transport systems determine the main principles of their formal description and modeling: decomposition, consideration of cause-and-effect relationships, consideration of integrity and hierarchy properties, consideration of the presence of free will, the possibility of choosing actions.

In this regard, when creating the theory of modeling of aviation transport systems, as a set of methods for solving practical problems, first of all, such problems as the mathematical description of the functioning of transport systems and its individual elements, information management in transport systems, decision-making processes in hierarchical active systems should be considered.

The basis of the mathematical apparatus used in this case can be the approaches of operations research, simulation modeling, game theory, queuing theory, information theory, decision theory, theory of active systems, etc. Along with a rigorous mathematical apparatus, so-called heuristic methods can also be used.

The efficiency of the aviation transport system is determined by the presence of active elements in its structure. Therefore, one of the tasks of the theory of modeling transport systems is a mathematical description, i.e. providing the possibility of quantifying the experience and intuition of a decision-maker in a hierarchical active system and systematization of methods for developing rational decision-making processes at its various levels.

When assessing the effectiveness of an aviation transport company, both a priori when it was created and when managing in the process of its operation, problems arise with a multi-criteria choice of a solution. Such problems are the essence of the theory and methods of decision-making, which develops methods for optimizing multi-criteria tasks in various situations and for various variants and cases of practice.

In this paper, attention is paid to such methods of constructing a quantitative model of decision-making processes that are most often used in practice, as the method of analytical hierarchy (MAH), used in solving so-called non-programmable tasks that require a creative approach.

In addition, in the process of creating, functioning and managing aviation transport systems, such issues as security are also relevant.

The economic recovery is inextricably linked with the development of vehicles and infrastructure of the transport system. The development of the infrastructure of aviation transport systems is associated with significant, global projects for the development of highways, airports and terminals, as well as communication, surveillance and management facilities in which one of the main requirements is to take into account safety conditions, risk assessment of dangerous situations. Such accounting is impossible without modeling transport processes. A wide range of the physical nature of such processes obliges to use the entire arsenal of models available today.

The paper pays special attention to current methods of studying aviation transport processes such as dynamic transport conditions, characteristics of transport space, operators and decision-makers.

Contents

Abbreviations

AC	Aircraft
AIS	Automated information system
AL	Airline
ATC	Air traffic control
ATS	Active transport system
CPS	Collision prevention system
CTC	Container terminal complex
DAS	Dynamic air situation
DM	Decision-maker
DMP	Decision-making process
DTS	Dynamic transport situation
HAS	Hierarchical active system
HF	Human factor
ILC	Information and logistics center
IS	Information support
PD	Professional development
PF	Personal factor
PT	Professional training
PTA	Professional thinking ability
SFA	Speech-functional act
TC	Transport company
TPM	Transport process management
TS	Traffic safety
TSM	Transport system management
TSp	Transport space
TSy	Transport system

Chapter 1
Transport Systems: Basic Concepts, Processes, Directions of Modeling and Their Research

1.1 Basic Concepts and Definitions

Transport is a branch of material production that transports passengers and cargo.

The production process is a sequential change in time of labor operations in the production of material goods, which act as production products.

Structural components of any production:

- the subject of labor;
- means (tools) of labor;
- production staff;
- production products.

The object of labor is an object of the production process, as a result of the technological transformation of which products intended for consumption are formed.

The subject of labor of transport production is transported goods and passengers.

The production process in transport is a change in the location of cargo and passengers in accordance with the needs of an individual, society, defense, industry and agricultural production of the country.

The products of the transport system are goods and passengers delivered to the final destination.

Means of production—industrial buildings, structures and equipment.

Means of transport production—transport space and transport equipment.

Transport space—zones of the Earth's surface, underground, water and air space equipped for moving and controlling the movement of transport objects.

The transport space includes:

- transport communications (railways, highways, waterways, air routes, pipelines);
- transport and technological terminals (loading and unloading and warehouse complexes, port and station facilities, complexes and buildings for customer service when placing an order for transport products);
- buildings, structures and cybernetic complexes of machines.

© The Author(s), under exclusive license to Springer Nature Singapore Pte Ltd. 2023

G. A. Kryzhanovsky et al., *Modeling of Transportation Aviation Processes*, Springer Aerospace Technology, https://doi.org/10.1007/978-981-19-7607-0_1

Transport equipment is a set of technical objects with the help of which the transport process is carried out.

Transport equipment includes:

- transport (mobile) vehicles that move goods and passengers through transport communications;
- equipment of transport and technological terminals intended for carrying out loading and unloading, transport and warehouse and intraterminal transporting technological operations;
- cybernetic machines designed for mechanization (automation) of management procedures in the process of customer service, vehicle management and management of organizational structures of transport.

Transport production personnel are people involved in transport and technological processes and management procedures in transport.

In accordance with the structure of the means of production, it is advisable to classify the personnel of transport production into groups:

- operators of production processes;
- programmers and operators of cybernetic machine complexes;
- managers and managers (managers) of organizational structures of transport.

1.2 Management Processes in Transport Systems

A system is an ordered set of objects (elements) united by any connections, the interaction of which is aimed at achieving a certain goal.

There is also such a definition: a system is a means to an end.

There are such concepts as system approach, system analysis, system analysis tools, weakly structured and unstructured management tasks, etc.

It can be said that system analysis is an interdisciplinary and supradisciplinary course that summarizes the methodology of the study of complex technical, natural and social systems and is the next step in the development of the general science of management.

The system approach considers an object not only in relation to other objects, but also as a system itself. The study of an object in external interactions makes it possible to determine the goals of its functioning, and the analysis of the internal structure makes it possible to assess the ways to achieve this goal.

Thus, system analysis is a discipline dealing with the problems of decision-making in conditions when the choice of an alternative requires the analysis of complex information of various physical nature.

A wide range of mathematical methods are used to analyze and synthesize complex systems.

The basis of the mathematical apparatus of system analysis is:

- linear and nonlinear programming;

- game theory;
- simulation modeling;
- queuing theory;
- theory of statistical inferences;
- theory of decision-making, etc.

When solving problems of system analysis, along with strict mathematical apparatus, so-called heuristic methods are also used.

The property of consistency is a universal property of matter. We can talk about the world as an infinite hierarchical system of systems. Moreover, parts of the system are in development, at different stages of development, at different levels of the system hierarchy and organization.

Every conscious action pursues a specific goal. In every action there are components, smaller actions. It is clear that these components should not be performed in an arbitrary order, but in a certain sequence. This approach to the construction of activities is called algorithmization.

Elements of creative activity (e.g., when solving managerial tasks) realized by a person "by intuition", in fact, can also be an unconscious realization of certain algorithmizable patterns, the realization of unconscious, but objectively existing and formalized criteria [1–4].

We can say that:

1. Every activity is algorithmic.
2. The algorithm of real activity is not always realized – a person performs a number of processes intuitively, i.e. his ability to solve some tasks is brought to automatism. This is a sign of professionalism, which does not mean that there is no algorithm in the actions of a professional.
3. In case of dissatisfaction with the result of the activity, the possible cause of failure should be sought in the imperfection of the algorithm.

In this regard, the identification and improvement of the algorithm is one of the ways to increase the consistency of activities.

The question of a scientific approach to the management of complex systems was raised by M. A. Ampere. He was the first to single out cybernetics as a special science of state management, designated its place among other sciences and formulated its systemic features.

The mass distribution of system representations is associated with the name of the American mathematician N. Wiener, who published the book "Cybernetics" in 1948 and then "Cybernetics and Society" N. Wiener and his followers pointed out that the subject of cybernetics is the study of systems.

Such scientists as Academician A. I. Berg, Academician A. N. Kolmogorov, F. I. Peregudov, F. P. Tarasenko made a significant contribution to the development of cybernetics.

Transport system—any organized association of objects of transport space, equipment and people for the implementation of transport activities.

The principal feature of transport systems in comparison with other production systems is that the transport process is carried out within a single transport space.

This determines the need for an orderly integration of individual transport systems into a well-formed organization.

An organization is an association of people, material, financial, energy and information resources so that their use ensures the achievement of a single goal of a combined set of objects.

The transport system has the following properties:

- integrity;
- hierarchy;
- activity.

The integrity property is due to the fact that the system has new properties compared to the properties of the individual elements of which it consists.

Organization implies the division of power and responsibility. In this regard, the hierarchy is a vertical division of power with the distribution of responsibility for a single result between individual structures.

The basis of hierarchical organization in transport is the functional division of power and responsibility. In this regard, there are:

- The federal system forming the upper level of the organization of transport;
- A territorial transport system representing the average level of transport organization;
- A production transport system that forms the basic level of transport organization.

The integration of objects into the system is carried out by means of connections, which manifest themselves in the form of information flows.

Information flow is a series of messages in the form of a set of signals and documents received by the system objects in a certain sequence.

Direct communication is information flows directed from the upper levels of the hierarchical ladder to the lower ones.

Feedback (response, reaction to a direct message)—information flows directed from the lower levels to the upper ones.

Coordinated and purposeful activity of many objects and subjects is carried out by means of transport system management.

Management is a process aimed at streamlining, maintaining security and improving the efficiency of the system.

The process is a consistent change in the state of the material world, in which certain patterns manifest themselves.

Quantitatively, any process manifests itself in changing the flows of energy, material bodies and information (all together or one of them).

The fundamental essence of the management process is the exchange of information between objects both within the transport system and between systems.

Thus, in the management process, the ordering, preservation and improvement of the efficiency of the transport system is carried out through the exchange of information between the three structures:

- object of management;
- system management;
- external objects.

The purpose of the transport system is to create transport products. Perturbations of the flows of information, energy, material resources (including the flow of customers, taking into account market conditions) and environmental impacts are imposed on the fulfillment of this goal. The scheme of interaction of elements of the transport system is shown in Fig. 1.1.

The effectiveness of the system is largely determined by the quality of decisions made daily by managers at different levels. In this regard, the task of improving decision-making processes (DMP) becomes important.

This problem is complicated by the multivariance of management tasks. A characteristic feature of the decision-making situation is the presence of a large number of possible options for action, from which you need to choose the best one.

For various objective reasons (in particular, due to the complexity of the tasks or the lack of sufficient necessary information), it is extremely rare to get the really best (optimal) option, therefore, the preferred, that is, the most appropriate option in these conditions, approaching the optimal in efficiency, is determined.

Management and planning is associated with the consideration of a more or less distant future and therefore always contains an element of uncertainty.

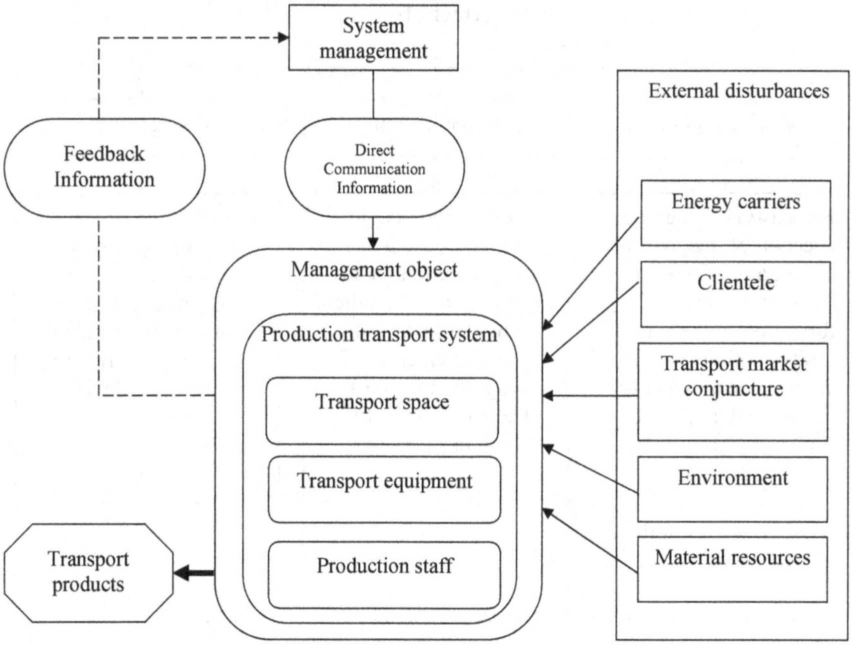

Fig. 1.1 Scheme of interaction between objects of the transport system

The head of the TSy must possess economic and mathematical methods, but he must also have intuition, sufficient practical experience, be able to use this experience, as well as the experience and knowledge of other specialists.

This is also required by such features of management tasks as the efficiency of the decision, insufficient information for decision-making, the need to take into account socio-psychological factors.

The management of the system is a set of subjects of the system that carry out management procedures that ensure the directed coordination of the activities of the management object to achieve its set goal.

The subjects (persons) who carry out management procedures are managers or managers. They, having information about the activities of the management object, implement their management functions.

Management functions are the actions of managers in programming, organization and management, ensuring the purposeful activity of all objects of the system (see Fig. 1.2).

Each manager implements the functions of the management according to his own algorithm, which, as already noted, means a sequence of operations (procedures) performed in a certain order.

In a generalized form, the algorithm of the management process for the implementation of management functions is reduced to the following procedures:

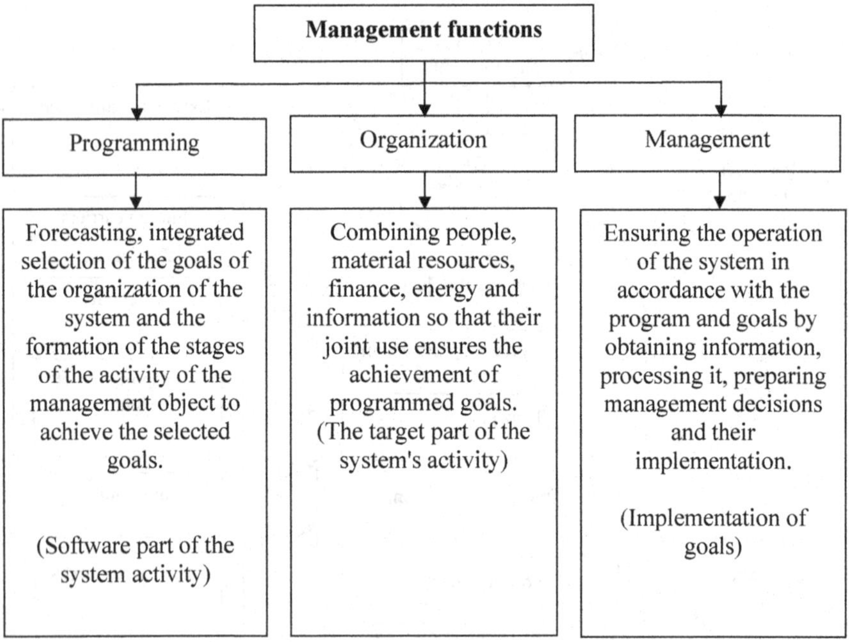

Fig. 1.2 Management functions

(1) Awareness of the general task of the transport system (TSy) and the general parameters of external disturbances;
(2) Receiving information in the feedback form about the parameters of the state of the transport system;
(3) Processing and analysis of the information received, on the basis of which the selection of goals and the formation of the program of actions of the management for the implementation of program activities is carried out;
(4) Organizational actions (including transformation of the organization structure) in order to concentrate resources on the implementation of program activities;
(5) Development of strategic and operational management functions in accordance with the program;
(6) Implementation of control actions by transmitting direct communication information to the control object;
(7) Control of the effectiveness of control actions on the control object based on the analysis of feedback information and correction of control actions.

1.3 Approaches to Modeling and Research of Transport Processes

The greatest difficulty in modeling and studying transport processes and systems is taking into account the presence in it:

(1) Active elements—a set of operators of technical means and transport space, managers in the organizational structure of the TSy;
(2) Uncertainties in the formation of programs and plans for the functioning of the TSy in the presence of competitive conditions;
(3) The hierarchical structure of the TSy;
(4) A fairly strict time limit on all processes, including: managerial and decision-making.

Therefore, when creating a theory of modeling transport processes and systems as a set of methods for solving practical problems, a number of problems arise:

I. Mathematical description of the functioning of the TSy and its individual elements to achieve the overall goal of the system. Explication of the goal—mathematical modeling of the TSy performance indicator—is not the simplest task in the problem of mathematical modeling.
II. Information support of management in the TSy, and especially in that part of it where there are decision-making processes (DMPs). Problems of "translating" information messages from the language of one level of the hierarchy to the language of another level;
III. The actual DMP in hierarchical active systems (HAS).

The property of activity is inherent in the so-called self-organizing systems. Self-organizing systems are systems that have the property of adapting to changing environmental conditions, capable of changing the structure when the system interacts with the environment, while maintaining the properties of integrity, systems that are able to form possible behaviors and choose the best ones [1].

These features are due to the presence of active elements in the structure of the system, which, on the one hand, provide the possibility of adaptation, adaptation of the system to new conditions of existence, on the other hand, introduce an element of uncertainty into the behavior of the system, which complicates the analysis of the system, the construction of its model, its formal description and, ultimately, complicate the management of such systems.

Therefore, one of the tasks of the theory of modeling of transport systems is concluded in a mathematical description, i.e. providing the possibility of quantitative expression of the experience and intuition of the decision-maker (DM) in a hierarchical active system with DMP and systematization of methods for the development of rational DMP at its various levels.

A typical example of the TSy-HAS is two airlines (AL) engaged in the transportation of passengers and cargo on long-haul flights (continental, interstate transportation). The structure of the hypothetical case of competition of such AL can be represented as in Fig. 1.3.

The managing elements in such HAS are AL representative offices in business centers or airports in the region. The fight is for passengers and freight.

The conjuncture on the world and regional markets, seasonal influences, weather, navigation period at any given time determine the state of the external environment, which, along with the state of the structure and the internal state of the HAS, synthesizes the general state of each of them.

The functioning of the HAS is a purposeful change in the state of the system over time.

The goals of the HAS are a given area in the HAS state space into which the state vector falls during the purposeful functioning of the HAS, e.g., the creation of transnational TSy, monopolization of transportation.

The tasks of the HAS are detailing, decomposition of the goal for different periods of its functioning. The tasks of the higher-level block elements in the hierarchy are the goals for the executive elements.

HAS management is the organization of its most rational functioning in order to achieve the goal in the "shortest possible" way. The control algorithm, as a rule, depends on the hierarchy level, but always includes a number of blocks, the main of which are (see Fig. 1.4):

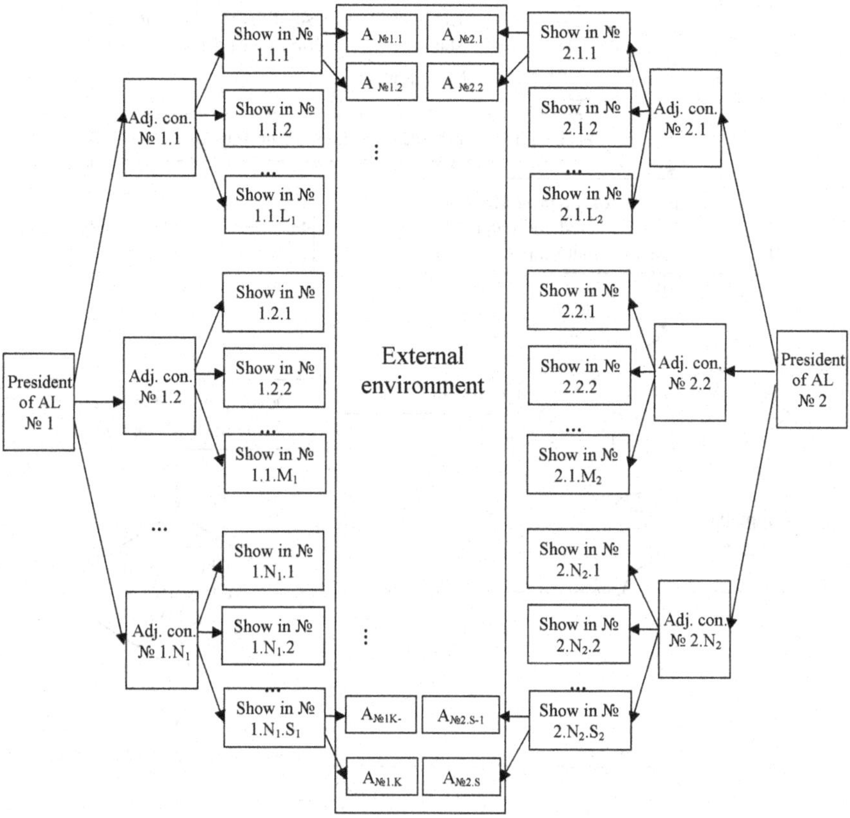

Fig. 1.3 Hypothetical structure of competing HAS-TSy-AL

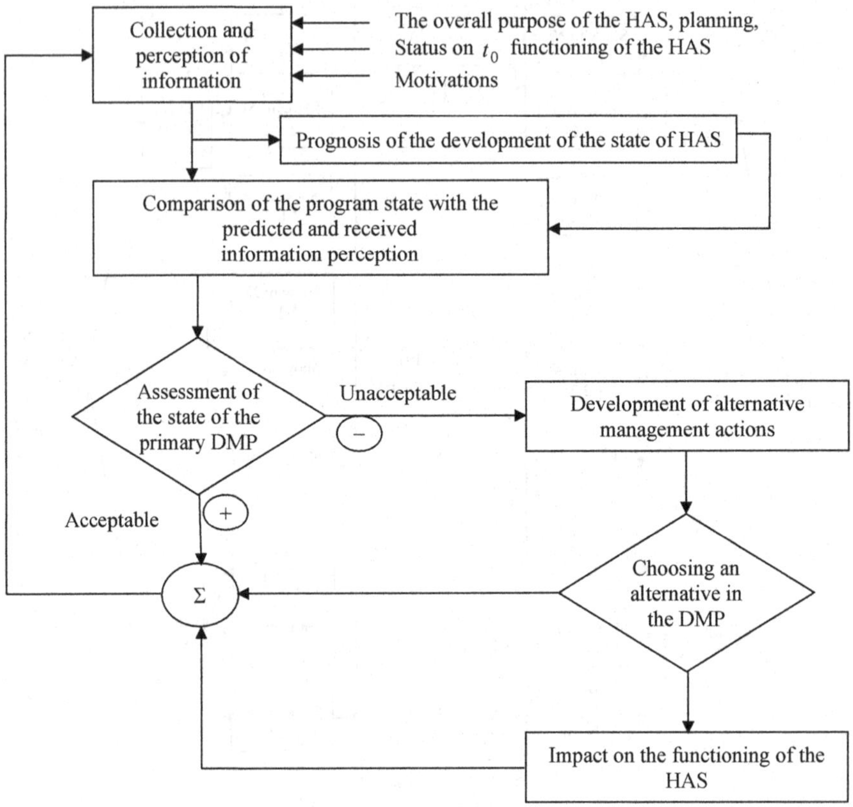

Fig. 1.4 Control algorithm in HAS

References

1. Burkov VN, Novikov DA (1999) Theory of active systems: state and prospects. Sinteg, Moscow, p 128
2. Galaburda VG, Persianov VA, Timoshinidr AA (1996) Unified transport system: studies for universities. In: Galaburdy VG (ed). Transport, Moscow, p 295
3. Kryzhanovsky GA, Kupin VV, Plyasovskikh AP (2008) Theory of transport systems. In: Kryzhanovsky GA (ed). GA University, St. Petersburg
4. Zaitsev EN, Bogdanov EV, Shaidurov IG, Pesterev EV (2008) General course of transport: a textbook for the study of discipline and the performance of control work. SPbGUGA, St. Petersburg, p 98

Chapter 2
Types of Models and Their General Characteristics—Principles of Modeling and Models of the Activity of Transport Companies as Hierarchical Active Systems (HAS)

2.1 The Role of Transport Process Modeling, Types of Models and Their Characteristics

Modeling is a common method of quantitative and qualitative (depending on the type of models) assessment of processes in any systems. It plays a special role in the study of transport processes.

Usually, a model is considered to be some, but necessarily simpler, similarity of the process under study. When solving the main task of air traffic control (ATC), in order to develop an optimal control command, the dispatcher must be able to predict the development of a dynamic air situation. One can imagine a device that would also be able to make such a forecast of the development of a dynamic air situation with no less accuracy. Such a device would serve as a model of the forecasting process, and only, since the dispatcher is able to perform other, much more complex functions, the repetition of which would require the introduction of other devices with an even more complex structure, etc.

Thus, the model is an approximate or simplified representation of the state and changes in the state of any particular process of the system under study and makes it possible to analyze and synthesize various processes of the system by common methods. The concept of a model is based on the presence of some adequacy, i.e. similarity, correspondence between the quantitative characteristics of two processes, one of which occurs in a real system, and the other in a model. If such adequacy is established in relation to the selected necessary quantitative or qualitative characteristics reflecting the state of the processes in any given sense, then it is said that there are relations between these processes of the original and the model, i.e. one of the processes is considered as the original, the other as a model. At the same time, adequacy or similarity, as well as the difference between the original and the model, can be used in various aspects. For example, you can only require the model to change (decrease) the geometric dimensions or the speed of the process.

At the same time, the concept of a simplified model is of great importance, which makes it possible to study processes of a complex nature with the help of models in

© The Author(s), under exclusive license to Springer Nature Singapore Pte Ltd. 2023

G. A. Kryzhanovsky et al., *Modeling of Transportation Aviation Processes*, Springer Aerospace Technology, https://doi.org/10.1007/978-981-19-7607-0_2

which only those features of the original that are essential for the range of phenomena studied are preserved. Such features can serve as external similarity when it comes to such types of models as layout, similarity of structure and, finally, one of the most important similarities is the similarity in the change in the state of processes described by a change in their quantitative characteristics.

In order to talk specifically about the correspondence of two processes or a process and its model, it is necessary to first introduce quantitative characteristics of the processes. Usually, they begin with the introduction of the most complete set of output parameters in terms of the studied properties and the smallest in terms of composition, by observing which one could judge the nature of the process. This is how the concept of a process state vector arises $X(t) = (x_1(t), \dots, x_n(t))'$. Each of the components of the state vector $x_s(t)$ characterizes one of the specific properties of the process. For example, when observing the process of movement of an aircraft, one can name $x_s(t)$—the longitudinal coordinate of the position of the aircraft relative to the Cartesian coordinate system, $x_2(t)$—its lateral coordinate, $x_3(t)$—height, $x_4(t)$—longitudinal, $x_5(t)$—lateral, $x_6(t)$—vertical components of the vector, flight speed and $x_7(t)$—fuel residue. The vector obtained in this case $X(t) = (x_1(t), \dots, x_7(t))'$ is often used to assess the condition of this aircraft in the general dynamic air environment of the ATC zone.

The components of the state vector in process modeling are to a certain extent analogous to generalized coordinates in classical mechanics. The analogy is confirmed even more fully if we talk about the state space, which is formed by components $x_s(t)$ considered as coordinate axes. The state of the process in such a space is then determined by a well-defined point. The general concept of the process state, which is currently most widely used in the study of complex systems, was introduced for the first time in 1936 by the English mathematical engineer Alan Matheson Turing, who created the theory of universal automata and the first universal computing machine in England. A similar approach was widely used in the study of problems of control theory and automation in 1942–1949 in the works of A. A. Andronov, A. I. Lurie, A. M. Letov, as well as in the 50 s in the USA mainly in the works of Richard Bellman. The most complete and final development of the concept of state space and the approach to the study of complex systems by state space methods is obtained in the modern theory of control and complex systems [1].

To assess the adequacy of the model and processes, it would be worthwhile, given a state vector, to assess the "similarity" of changes in this vector under the same conditions. As such conditions, one can, for example, consider external influences, control signals acting on the process and, accordingly, on the model, etc. One can imagine a line with a scale of "similarity", where the complete absolute coincidence of changes in the vector of the state of the model and the process serves as the beginning, and other provisions determine different degrees of increasing difference under the same initial conditions. This is how such a type of variables as input signals is defined, generalizing external conditions in the form of an input vector.

Comparison of the model and the real process can now be carried out using the introduced concepts of the state space method (Fig. 2.1). With such a comparison, the

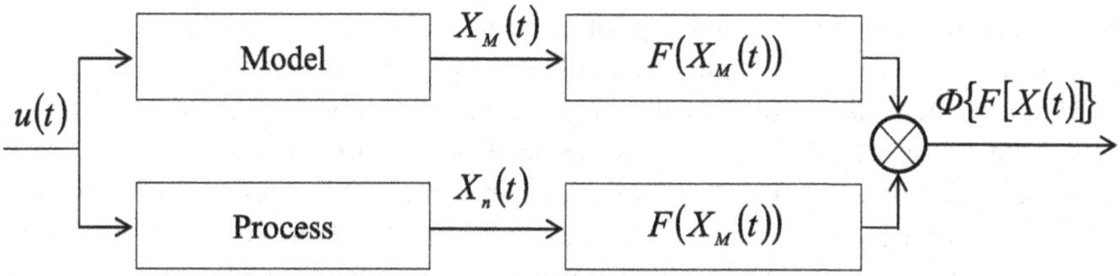

Fig. 2.1 Comparison of the model and the real process

model is called isomorphic if there is a complete correspondence between the model and the original, i.e. the real process. To clarify the concept of "full compliance", consider the case when the state of a process is determined by its input signals $u(t) = (u_1(t), ..., u_m(t))$ and output signals $X(t) = (x_1(t), ..., x_n(t))'$, which ultimately determine the quantitative characteristics of the process.

If two processes are characterized by identical sets of input and output signals and the change in their states $X_M(t)$ is the same when the input changes $X(t)$, then they are said to be isomorphism or full compliance. Isomorphic processes are indistinguishable from each other when observing only the input and output components of their vectors $u(t)$ and $X(t)$. However, in many cases isomorphic models turn out to be extremely complex and inconvenient for practical use due to the fact that the number of components of the vector $X_M(t)$ turns out to be large. Isomorphism conditions are not necessary conditions for the model to correspond to the original in all cases of studying the properties of the process.

Often the model can also be used when its correspondence to the original is not as complete as the isomorphism conditions require. So, for example, if among the quantitative characteristics of a process that determine its state vector, there are significant and less significant characteristics in relation to the task being studied in this process, then it is possible to require the model to correspond to the original only with respect to the essential ones. Then, instead of the original with the dimension of the state space n, a simplified model with the dimension of the state space k < n is obtained. Simplification in the model can be obtained not only by reducing the number of components of the state vector, but also by combining a certain set of states into one, i.e. as a result of a rough consideration of the state change over time. So sometimes a dynamic process with a relatively short transition from one state to another can be studied using a static model that is adequate to the original in its final state.

A process learned from the original process in a real system by simplifying by reducing the dimension of the state vector or a more rough estimate of the change in its state over time is called a homomorphic or simplified model of the original process. We can say that the homomorphic model contains elements corresponding only to large parts of the original, but there is no complete element-by-element correspondence between them. The relationship between the original and its model is unequal, since the original cannot be considered homomorphic for the model (unlike the isomorphic case). As an example, you can again use the process of controlling the movement of an aircraft. For the pilot, the essential values are all the components of the velocity vector of the center of mass of the aircraft and its position in a given coordinate system, the values of the angles and angular velocities of the aircraft around the center of mass, as well as the position of all controls and many other parameters. The dispatcher of the district center of the ATC system, which controls the air traffic of a group of aircraft in this zone, is interested only in the components of the position of the center of mass of a group of aircraft and (to a lesser extent) the vector of their linear velocity. Such a replacement of the original state space with respect to a single aircraft (including a set of values of all its components) with a space consisting of four to six components means a transition to a homomorphic model.

The above arguments concerned, so to speak, the ideological side of modeling. In addition, there is a formal or external side of modeling, concerning the form and type of models. Currently, three types of models are widely used in the study of complex systems: flowcharts, or graphs of systems, mathematical models and physical models. Flowcharts or graphs are sometimes called "iconographic" models. They are used to represent the structure of systems and their corresponding functional relationships. Examples of iconographic models are schemes (Figs. 1.1 and 1.2) structure (Fig. 1.3.), a block diagram (Fig. 1.4.), and also for the ATC system is a block diagram of a set of the simplest control circuits aircraft pilot—radio communication device—dispatcher—radar—aircraft. The arrows in the flowchart show the directions of signal flows that control processes or carry information about the state of processes in the system. It is clear that such models give only a structural representation in the system. For a more complete study, a physical or mathematical model is more suitable.

The essence of physical models consists in using the results of formal similarity between some components of processes occurring in real systems and homomorphic models that differ from the original in their physical nature and general structure. Examples of physical modeling can serve as electrical analogues of mechanical or pneumohydraulic systems. In transport systems, for example, in ATC systems, physical modeling is used in the construction of simulators and simulators. The most complete picture of the dependence of the output characteristics of the system on the input can usually be obtained using mathematical modeling of processes.

Mathematical models reflect the dependencies between the inputs and outputs of the system by means of graphs, tabular values, computer programs, equations, systems of equations or inequalities indicating the permissible ranges of magnitude variation. In the presence of a complete mathematical model, it is a convenient means of analyzing the processes occurring in a real system. The mathematical model of the system is usually considered to be descriptions in some formal language of changes in the most significant components of the quantitative characteristics of the state of the process. With the help of formal procedures over such descriptions, it is possible to obtain a number of conclusions and judgments about the features of the process under study that interest the researcher. This is the meaning of analyzing the processes of systems using mathematical modeling. Further, wherever we are talking about models, mathematical models are most often meant. Mathematical models cannot be comprehensive and ideally reflect changes in the state of the process; they do not describe a real system, but its homomorphic, simplified model.

Usually, the construction of a mathematical model is a procedure that does not follow any particular unchanging example or pattern. Here we can only talk about uniform requirements for the model. An essential requirement is the proximity of the specified characteristics of the model and the original. This requirement can sometimes be expressed by introducing a generalized characteristic (e.g., a Lyapunov function, or a model performance indicator). Currently, two approaches are used in mathematical modeling of transport processes, according to which process models are considered either as characteristics of discrete control objects, such as aircraft, or as characteristics of continuous traffic flows formed by aggregates of incoming, flying and departing aircraft. In the first of the cases under consideration, a state vector is introduced and it becomes possible to quantify the dynamic air situation in the ATC zone. The second also uses quantitative characteristics, such as intensity, density of air traffic and others, i.e. functions are introduced that describe the change in the state of the flight process not of one aircraft, but of some combination of them forming a flow. As in both cases, when modeling, we are talking about the proximity of the model and the real process. If the process is characterized by a quantitative measure—the measured values of the components of the state vector $X(t) = (X_1(t), \ldots, X_n(t))$ or "continuous" characteristics in the form of functions describing the state of the flow of aircraft, and the mathematical model is given in the form of relations determined by arguments, as which the components of the state vector or continuous characteristics (intensity) are taken flow, then the model is suitable only if the conditions are met:

$$\begin{aligned} &\Phi_1\{ F_{(1)}[X(t)] - F_{(1)}[X_M(t)] \,|\,\} \leq \varepsilon_1(t, X); \\ &\Phi_2\{ F_{(2)}[\lambda(t)] - F_{(2)}[\lambda_M(t)] \,|\,\} \leq \varepsilon_1(t, \lambda); \end{aligned} \tag{2.1}$$

where $\varepsilon_{1,2}$ is a predetermined function of time and a component of the vector of the state of the process or the intensity of the flow; $F_{1,2}$—a function that is a generalized characteristic of the process; $\Phi_{(1,2)}$—functional, i.e. the function of $F_{(1,2)}$ characterizing the proximity of the process and the model.

Another requirement for a mathematical model is, perhaps, its greater simplicity. Due to the contradictory requirements, the construction of a model often involves, in addition to a scientific approach, the manifestation of some art or intuition.

The task of constructing a mathematical model can be formulated as an optimization problem—to find a model acceptable under condition (2.1), the simplicity of which would be the greatest. It can be solved only if the simplicity requirement is formalized in the form of a constructive performance indicator. Taking into account the complexity of constructing mathematical models of transport processes, it is necessary to understand the advantages due to which mathematical modeling necessary for the use of transport processes has become one of the main methods of analysis and improvement. These primarily include the following:

- the presence of mathematical models of processes allows numerical experimentation in the study of their characteristics, which guarantees complete safety of the system processes;
- the time scales in numerical studies of process models can be significantly less than the real time scale of the actual process, which speeds up research and allows you to obtain a number of qualitative results;
- reproducibility of the study of mathematical models allows you to establish the causes and prerequisites of unexpected results, which is often impossible in a real system;
- conducting experiments using mathematical models is economically more profitable than experimenting on real processes in the system.

These advantages served as a justification for the fairly widespread introduction of methods of mathematical modeling of transport processes not only in the design of transport systems, but also for the purpose of analyzing the quality of functioning of individual elements, making recommendations on the reorganization of the structure of systems, traffic management issues and many others.

2.2 Principles of Modeling Processes in HAS

The need to study the general functioning of the HAS forces us to search for and use various types of mathematical descriptions of structural schemes, information, energy and mass transfer in them. To use the entire arsenal of mathematical models, it is necessary to detail the functioning (activity) HAS in the form of functions of its individual levels, blocks of each of the levels and elements on the lowest of them.

This is how the first principle of formal description—modeling of HAS arises—the principle of decomposition, one of the general principles of the study of large systems. According to this principle, the modeling of the functioning of the HAS should begin with the modeling of individual processes of the functioning of the elements.

For a transport system, each of these elements is a set or a separate vehicle (Ve), forming together with a part of the transport space a certain image, a dynamic transport situation (situation) (DTS).

DTS is a consequence of the processes of Ve movement, the processes of changes in the transport space (changes in equipment, the external environment, etc.). All these changes are subject to causal relationships and patterns.

Consideration of causal relationships is the second principle of formalization of modeling HAS. The presence of various kinds of uncertainties and random factors in the functioning of the HAS leads to the need to use an empirical approach, i.e. along with known patterns and relationships, the empirical level is also actively used in the modeling of the HAS.

Since any transport system can be represented as multilevel, the third principle in its modeling is precisely the need to take into account the properties of integrity and hierarchy.

As is known, the integrity of the HAS consists in the impossibility of presenting its characteristics in the form of a simple aggregate (sum, for example) characteristics of individual elements. This property is sometimes associated with the property of aggregation: The characteristics of complex systems are not a property of its individual elements having different physical nature.

For the HAS, this means that modeling the functioning of each of the levels of its hierarchy should include a certain set of parameters that characterize the integral property of the entire HAS.

The hierarchy of any transport system means that one of the conditions for its modeling is the choice of parameters that characterize it as:

- functioning in a more general system (the economy of a region, country or the world as a whole);
- a complete system in itself;
- a complex structure that includes many elements of different physical nature.

Thus, the level of description (modeling) is determined by the specific task of research and the level of analysis of processes in the HAS, and the parameters selected for this should have a clear meaningful meaning and a clear physical (technical) meaning.

The presence of personalities (ergatic elements) and collectives in the HAS structure determines its active properties. Taking into account the presence of free will, the possibility of choosing actions, including counteraction in a competitive struggle, is the fourth principle of modeling HAS and such prominent representatives of systems of this class as transport systems.

Thus, for a comprehensive study of the HAS, it is necessary to have a set of mathematical models reflecting its main properties: decomposition, causal relationships, hierarchy, integrity and activity.

2.3 Modeling the Functioning of the HAS

Let's consider the typical structure of two competing transport systems (Fig. 1.3)—transport companies (air, sea, river, etc.). Denote by $i = \overline{1, N_i}$ the numbers of elements in the transport company N_1 and by $j = \overline{1, N_j}$—in the transport company N_2, where N_i, N_j is the number of elements in each of them, it is possible to determine (denote, explicate) the relationship (relationship) of any two elements of the structure using binary (0, 1) vectors with components characterizing the types of relationship (informational, financial, energy or turnover, etc.):

$$a_{\text{ij}} = \begin{pmatrix} a_{\text{ij}}^{(I)} \\ a_{\text{ij}}^{(\text{u})} \\ a_{\text{ij}}^{(\Phi)} \\ \vdots \end{pmatrix}; \overline{a}_{\text{ik}} = \begin{pmatrix} \overline{a}_{\text{ik}}^{(I)} \\ \overline{a}_{\text{ik}}^{(\text{u})} \\ \overline{a}_{\text{ik}}^{(\Phi)} \\ \vdots \end{pmatrix}$$

The following properties of the relations of elements in the hierarchical structure are accepted here:

$$\begin{aligned} & a_{\text{ijk}} \subset a_{\text{ij}} \subset a_i \subset a, \\ & a = \bigcap_{i=1}^{N_i} a_i \forall_i a_i \bigcup_{j=1}^{N_j} a_{\text{ij}}, \ldots, \\ & \bigcap_i a_i = 0, \ \bigcap_j a_{\text{ij}} = 0. \end{aligned} \tag{2.2}$$

That is a_{ij}^{u}—means the presence ($a = 1$) or absence ($a = 0$) of a control connection (u) between elements i and j—direct and, if a_{ji}—feedback, if all this is inside the transport company N_1.

All the same, but with a line—for relations between elements of companies N_1 and N_2: $\overline{a}_{\text{ik}}^{I}$—information exchange between the i-th element of various companies N_1 and the k-th element of the company N_2.

Thus, competition is represented (explicated) in the form of the presence of connections $\overline{a}_{\text{ik}}$, for example, in the market $\overline{a}_{\text{ik}}^{p}$. So the information connection of the i-th element N_1 with the k-th element of the transport system N_2 is denoted as $\overline{a}_{\text{ik}}^{(I)}$, and accordingly, $\overline{a}_{\text{ki}}^{(I)}$ the feedback of the k-th element with the i-th. Elements form a structural (conditionally) matrix of a transport company at this level of research of transport systems:

$$C_{N_1} = \begin{vmatrix} 0 \ a_{12} \ a_{13} & \ldots & a_{1k} & \ldots & a_{1N_1} \\ \ldots \ 0 \ \ldots & \ldots & \ldots & \ldots & \ldots \\ \ldots \ \ldots \ \ldots & \ldots & \ldots & \ldots & \ldots \\ a_{k1} \ a_{k2} \ a_{k3} & \ldots & 0 & \ldots & a_{kN_1} \\ \ldots \ \ldots \ \ldots & \ldots & \ldots & \ldots & \ldots \\ a_{N_1 1} \ a_{N_1 2} \ a_{N_1 3} & \ldots & a_{N_1 k} & \ldots & 0 \end{vmatrix}$$

Similarly, matrices of type C_{N_2} are constructed, as well as matrices of type $C_{N_1 N_2}$ and $C_{N_2 N_1}$, which characterize the influence of company N_1 on company N_2 and vice versa, respectively.

The competition of companies forms a structure—a matrix of the structure of the state of the system:

$$C = \left| \begin{array}{c|c} C_{N_1} & C_{N_1 N_2} \\ \hline C_{N_2 N_1} & C_{N_2} \end{array} \right| \tag{2.3}$$

Thus, matrix C fully characterizes the structure of interaction and communication, both within each of the companies and between them, i.e. it is a matrix of the state of the system structure.

In addition to the state of any transport system, the most important characteristic is the position of its elements (vehicles, representative offices) in space (e.g., in geographical coordinates λ, ϕ, γ).

The relative position of the elements can be set by the vector $\rho_{\text{ij}}(t)$: $\rho_{\text{ij}} = \rho_{\text{ij}}\left(\rho_i^{(t)}, \rho_j^{(t)}\right)$, then, by analogy with (2.3), a matrix of the spatial state of the system can be set:

$$R^{(t)} = \left| \begin{array}{c|c} R_{N_1}^{(t)} & R_{N_1 N_2}^{(t)} \\ \hline R_{N_2 N_1}^{(t)} & R_{N_2}^{(t)} \end{array} \right|, \tag{2.4}$$

where $R_{N_1 N_2}^{(t)}$ is the matrix of the relative position of the elements of the transport system N_1 relative to the elements of the transport system N_2.

By introducing, by analogy, vectors of the state of the external environment, dividing it "by geography" of companies, we get:

$$W(t) = \left| \frac{W_{N_1}}{W_{N_2}} \right| = \left| \begin{array}{c} \omega_1 \\ \vdots \\ \omega_{N_1} \\ \omega_k \\ \vdots \\ \omega_{N_2} \end{array} \right|. \tag{2.5}$$

Similarly, for the vectors of the internal state of the elements of each of the companies (vehicle speed, cargo and passenger capacity, costs, the presence of control from representatives or federal agencies, the availability of loans, etc.—for each element), you can write:

$$P(t) = \frac{P_{N_1}}{P_{N_2}} = \frac{\begin{pmatrix} p_1 \\ \vdots \\ p_{N_1} \end{pmatrix}}{\begin{pmatrix} p_k \\ \vdots \\ p_{N_2} \end{pmatrix}}. \tag{2.6}$$

So, the introduced variables in their totality fully characterize the state of each of the companies and, moreover, their interaction. Then we can imagine a certain state space defined by the values of the components of the vector (matrix) of the system N_1—N_2:

$$\begin{aligned} S(t) &= \{C(t), R(t), W(t), P(t)\} \\ &= \left\{ \begin{vmatrix} C_{N_1} & C_{N_1N_2} \\ C_{N_2N_1} & C_{N_2} \end{vmatrix} \begin{vmatrix} R_{N_1} & R_{N_1N_2} \\ R_{N_2N_1} & R_{N_2} \end{vmatrix} \begin{vmatrix} W_{N_1} & \\ & W_{N_2} \end{vmatrix} \begin{vmatrix} P_{N_1} & \\ & P_{N_2} \end{vmatrix} \right\} = \begin{vmatrix} S_{N_1} & \\ & S_{N_2} \end{vmatrix} \\ &= \begin{vmatrix} s_1 & & & & \\ & \ddots & & & \\ & & s_i & & \\ & & & \ddots & \\ & & & & s_j \\ & & & & & \ddots \end{vmatrix}; \quad \begin{aligned} S_{N_1} &= \left\{C_{N_1}, C_{N_1N_2}, R_{N_1}, R_{N_1N_2}, W_{N_1}, P_{N_1}\right\}; \\ S_{N_2} &= \left\{C_{N_2}, C_{N_2N_1}, R_{N_2}, R_{N_2N_1}, W_{N_2}, P_{N_2}\right\}, \end{aligned} \end{aligned} \tag{2.7}$$

where S_{N_1}, S_{N_2} are the states of competing companies.

The element of the state vector of the system is defined as

$$s_i = \{C_i, R_i, W_i, P_i\}, \tag{2.8}$$

where C_i is the vector of the connection of the i-th element with other elements of the system, i.e. the i-th row of direct connections and the j-th column of feedbacks of the matrix C (2.2); is the vector of the position of the i-th element, etc.

Note that the vector $S(t)$ in (2.7) is a complete formal description of the state of the system at time t for any level of its detail or generalization.

2.4 Simulation of Changes in the State of the System Over Time

The functioning of the system—i.e. the change in the state of the system in time is determined discretely—in steps-cycles and the trajectory of the state vector depends on the control vector

$$S(n) = S\{U(n)\}: \tag{2.9}$$

$$U = \begin{vmatrix} \mathrm{u}_1 \\ \vdots \\ \mathrm{u}_i \\ \bar{\mathrm{u}}_1 \\ \vdots \\ \bar{\mathrm{u}}_j \end{vmatrix} = \begin{vmatrix} U_{N_1} \\ U_{N_2} \end{vmatrix} \tag{2.10}$$

Indeed, at each $(n+1)$-th step, the elements of the system interact with each other and change their state under the influence of controls $P(n+1) = \{P(n), U(n)\}$, which in turn leads to a change in the position $R(n+1)$, environment $W(n+1)$, structure $C(n+1)$, i.e. the entire system (2.7) in accordance with the causal principle.

Now the step-by-step dynamic process of changing the state during system management can be represented by a sequence of operators of the form:

$$F_S : \begin{cases} F_p : \{P(n), C(n), R(n), W(n), U(n)\} \Rightarrow (m.e.\, //npu\beta o\partial um\ \kappa//)\ P(n+1) \\ F_R : \{P(n+1), C(n), R(n), W(n), U(n)\} \Rightarrow R(n+1) \\ F_W : \{R(n+1), W(n)\} \Rightarrow W(n+1) \\ F_C : \{P(n+1), C(n), R(n+1), W(n+1), U(n)\} \Rightarrow C(n+1) \\ F_U : \{P(n+1), C(n+1), R(n+1), W(n+1), U(n)\} \Rightarrow U(n+1) \end{cases} \tag{2.11}$$

that is, an operator is a step-by-step sequential process of applying operators F_P, F_R, F_W, F_C, F_U.

It can be argued that the most interesting is a detailed consideration of the operator describing the interactions of the elements of the system and representing a matrix of operators F_P of transformations of elements

$$\begin{vmatrix} \Pi_1 \cdots \cdots \cdots & & & & & \\ & \ddots \cdots \cdots & & & & \\ & \Pi_i & \cdots \Pi_{i\mu} & \overline{\Pi}_{ij} & \cdots & \overline{\Pi}_{i\xi} \\ & \vdots & \ddots \quad \vdots & \vdots & & \vdots \\ & \Pi_{\mu i} & \cdots \Pi_{\mu} & \overline{\Pi}_{\mu j} & \cdots & \overline{\Pi}_{\mu\xi} \\ & & & \ddots & & \\ & & & \Pi_j & & \\ & & & & & \Pi_{\xi} \end{vmatrix} = \begin{vmatrix} \Pi_{N_1} & \Pi_{N_1 N_2} \\ \Pi_{N_2 N_1} & \Pi_{N_2} \end{vmatrix}, \tag{2.12}$$

where Π_i is the consequences operator that displays changes in the internal state of the i-th element as a function of the impact of other elements in the previous step, for example, the restoration of its financial position after damage caused by the actions of a competitor; $\Pi_{\mu i}$, $\Pi_{\xi j}$ are the consequences operators for elements within companies N_1 and N_2; $\overline{\Pi}_{ij}$, $\overline{\Pi}_{ji}$ are the operators of mutual consequences of companies, respectively, as a result of the influence of the i-th the company element N_1 on the j-th element of the company N_2 and the j-th element of the company N_2 on the i-th element of the company N_1.

Let the impact of the elements of your company on the i-th element be described by a function f_1, the impact of the same element on other elements of your company—by a function f_2, the impact of elements of a competing company—ϕ_1 and the impact of this element on elements of a competing company—ϕ_2. Then the changes in the internal state of any i-th element at the $(n+1)$th step can be described:

$$\begin{aligned} P_i(n+1) = F_p^{(i)}\big[&P_i(n),\, f_1\big(\{\Pi_{\mu i}(n)\cdot a_{\mu i}(n)\}\big),\, f_2\big(\{\Pi_{i\mu}(n)\cdot a_{i\mu}(n)\}\big),\\ &\phi_2\big(\{\overline{\Pi}_{ij}(n)\cdot a_{ij}(n)\}\big),\, \phi_1\big(\{\overline{\Pi}_{ji}(n)\cdot a_{ji}(n)\}\big),\, \Pi_i(n)\big], \end{aligned} \tag{2.13}$$

where

$\Pi_{\mu i}(n) = \Pi_{\mu i}\big[P_i(n), P_{\mu i}(n), \rho_{\mu i}(n), u_{\mu i}(n), W_\mu(n)\big]$ for all $\mu = \overline{1, N_2}$;
$\Pi_{i\mu}(n) = \Pi_{i\mu}\big[P_\mu(n), P_{i\mu}(n), \rho_{i\mu}(n), u_{i\mu}(n), W_i(n)\big]$ for all $i = \overline{1, N_1}$;
$\overline{\Pi}_{ij}(n) = \overline{\Pi}_{ij}\big[P_i(n), P_j(n), P_{ij}(n), \rho_{ij}(n), u_{ij}(n), W_i(n)\big]$ for all $i = \overline{1, N_1}$;
$\overline{\Pi}_{ji}(n) = \overline{\Pi}_{ji}\big[P_j(n), P_i(n), P_{ji}(n), \rho_{ji}(n), u_{ji}(n), W_j(n)\big]$ for all $j = \overline{1, N_2}$;

$$\Pi_i(n) = \Pi_i[P_i(n), \rho_i(n), W_i(n)].$$

Here it is assumed that in the course of competition, interacting elements spend or receive resources from other elements of their company and, when exposed to elements of a competing company, lose part of their resources.

So, the time characteristics of the functioning of the elements and the entire system as a whole remain uncertain. It is clear that depending on the subject orientation of the research, these characteristics may be different. However, in any transport system and in the HAS, it is possible to identify a number of clear states of elements and determine the time cycles of their functioning.

Indeed, let a period of time pass from the beginning of the receipt of the command (control) on the element to the beginning of its (i-th element) actions:

$$\tau_i^{(1)}(n) = \tau_i^{(1)}[P_i(n), W_i(n)]. \tag{2.14}$$

The very action of the element—the execution of the task by it requires a period of time

$$\tau_i^{(2)}(n) = \tau_i^{(2)}[P_i(n), U_i(n), W_i(n)] \le \tau_{i\,\max}^{(2)}. \tag{2.15}$$

And finally, the return of the element to the working state after the task is completed:

$$\tau_i^{(3)}(n) = \tau_i^{(3)}[P_i(n), W_i(n)] \ge \tau_{i\,\min}^{(3)},\ \text{where}\ i = \overline{1, N_1}. \tag{2.16}$$

Then the cycle of functioning of the i-th element

$$T_i(n) = \sum\nolimits_{\nu=1}^{3} \tau_i^{\nu}(n). \tag{2.17}$$

The cyclicity of the i-th element, i.e. its being in a working or non-working (already occupied or out of order) state, can be set using the Kronecker symbol

$$k_i(n) = \begin{cases} 1 \text{ by } \tilde{T}_i^{(r)}(n) \le \tilde{T}_i^{(\max)}(n), \\ 0 \text{ if } T_i(n) \ge T_i^{(\min)}(n), \end{cases} \tag{2.18}$$

$\tilde{T}^{(r)} = \tau_i^{(1)}(n) + r\tau_i^{(2)}(n) + r\tau_i^{(3)}(n); r = 0 \div K+1; K+1$—number of cycles, that $k_i(n) = 1$ is, it means that the ith element is ready to act at the n-th step, while the $k_i(n) = 0$ i-th element at step n is busy and/or does not function.

Then the cyclicity matrix of the system K can be represented for all elements of the system

$$K: \{P(n+1), C(n+1), R(n+1), W(n+1), U(n+1)\};\ K(n) \Rightarrow K(n+1), \tag{2.19}$$

$$K = \begin{vmatrix} k_i^{(N_1)} & \\ & k_j^{(N_2)} \end{vmatrix} \tag{2.20}$$

and changing the state of the i-th element now instead of (2.13) will represent a dependency of the form:

$$\begin{aligned} P_i(n+1) = F_p^{(i)}\big[&P_i(n), f_1\big(\{\Pi_{\mu i}(n) \cdot a_{\mu i}(n) \cdot k_\mu(n)\}\big), \\ &f_2\big(\{\Pi_{i\mu}(n) \cdot a_{i\mu}(n) \cdot k_i(n)\}\big), \phi_2\big(\{\bar{\Pi}_{ij}(n) \cdot a_{ij}(n) \cdot k_i(n)\}\big), \\ &\phi_1\Big(\Big\{\bar{\Pi}_{ji}(n) \cdot a_{ji}(n) \cdot k_j^{(N_2)}(n)\Big\}\Big), \Pi_i(n)\Big]. \end{aligned} \tag{2.21}$$

The internal state of the i-th element is determined by the dependency

$$P_i(n) = P_i[U(n), \ldots]. \tag{2.22}$$

Thus, dependencies (2.12, 2.13, 2.21 and 2.22) determine the state of the i-th element at the $(n+1)$-th step, depending on the internal state at the n-th step, the effects on it of the elements of the company N_1 and the competitor company N_2, the structure of connections and their cycles works.

It is clear that during the company's activity, during the execution of the cycles of the i-th element (e.g., a vehicle), its position in the space environment of the system's activity changes. Then these changes can be represented by the equations:

$$\left.\begin{array}{c}\rho_i(n+1)=\rho_i(\rho_i(n),V_i(n+1))\\ V_i(n+1)=V_i\Big\{V^{(0)},P_i(n),U(n),\Big[P_j^{N_2}(n),U_j^{N_2}(n),a_{\mathrm{ji}}^{N_2}(n)\Big],W(n)\Big\}\\ j=\overline{1,N_2};\,i=\overline{1,N_1};\\ \rho_{\mathrm{ij}}(n+1)=\rho_{\mathrm{ij}}\big[\rho_i(n+1),\rho_j(n+1)\big].\end{array}\right\} \quad (2.23)$$

where V is the vector of the rate of change of the coordinates of the i-th element in the medium, depending on $V^{(0)}$ the possible maximum velocity; $P_i(n)$—the state of the element, control $U(n)$, counteraction of the competitor, if $a_{\mathrm{ij}}^{N_2}(n) \neq 0$ and the state of the environment $W(n)$.

The natural environment also changes when the i-th element is moved

$$W_i(n+1) = W_i[\rho_i(n+1)], \quad (2.24)$$

for the system as a whole:

$$W(n+1) \Rightarrow W\{R(n+1), W(n), P(n+1)\}. \quad (2.25)$$

Thus, all the dependencies characterizing changes in the state of the system as a whole during management and cycles of changes in the temporal structure of its functioning are obtained. To do this, you need to make changes to each line of the dependency system (2.11) in the form (2.21), (2.23), (2.25) and so on, and add another one to it:

$$F_K : \{P(n+1), C(n+1), R(n+1), W(n+1), U(n+1), K(n)\} \Rightarrow K(n+1), \quad (2.26)$$

$$F_\nu, \ \nu = P, R, W, C, U.$$

Note that the modeling of the transport activity of the HAS in the form of dependencies (2.26) meets the requirements of all principles (decomposition, causality, hierarchy, activity) when describing the system. Any actions of competing and opposing transport companies can be specified when determining the type of functions f_ν and ϕ_ν, F_ν in the generalized system of Eqs. (2.26).

Reference

1. Kryzhanovsky GA, Kupin VV, Plyasovskikh AP (2008) Theory of transport systems. In: Kryzhanovsky GA (ed). GA University, St. Petersburg

Chapter 3
Fundamentals of Modeling Control Processes in Transport Systems—HAS

3.1 Management Processes in the HAS. Decision-Making Processes and Their Modeling

The experience of studying transport systems of various types allows us to conclude that there is a typical hierarchical structure of their management, including four levels of management: the federal state level of regulation of transport activities, the departmental level and the level of traffic management of this type of transport, the level of the transport campaign and, finally, the level of direct operational control of the vehicle and, in relation to the transport space, the level of control of their movement (Fig. 3.1).

As it is obvious, this scheme as an iconographic model does not contradict the classification scheme of transport production personnel, but only details the relationships between individual blocks of the scheme and information flows between them. At the same time, it is clear that the main activity of the personnel of the last two levels—the level of the transport company and its managing staff together with a group of research consultants—consists in developing and making decisions. This type of activity occupies a special place in the modeling and research of processes in the HAS.

Decision-making is a special type of activity of the operator and/or manager, consisting in choosing one of several solutions. The peculiarity is that in the decision-making process (DMP), significant intellectual effort—costs are required and, what is the peculiarity—volitional (mental strength) effort—costs.

The person who carries out the decision-making process in the management structure of transport systems is called the decision-maker (DM).

Since the safety of passengers and crew of vehicles, as well as the economic indicators of the transport system (TSy), often depends on the choice, i.e. on the DMP, it becomes clear that the methods that allow finding the best optimal solutions in a given time, sometimes turn out to be simply invaluable.

At the same time, the DMP processes include both the generation of indicators and the selection, type, their convolution and the generation of solutions, not to mention

© The Author(s), under exclusive license to Springer Nature Singapore Pte Ltd. 2023

G. A. Kryzhanovsky et al., *Modeling of Transportation Aviation Processes*, Springer Aerospace Technology, https://doi.org/10.1007/978-981-19-7607-0_3

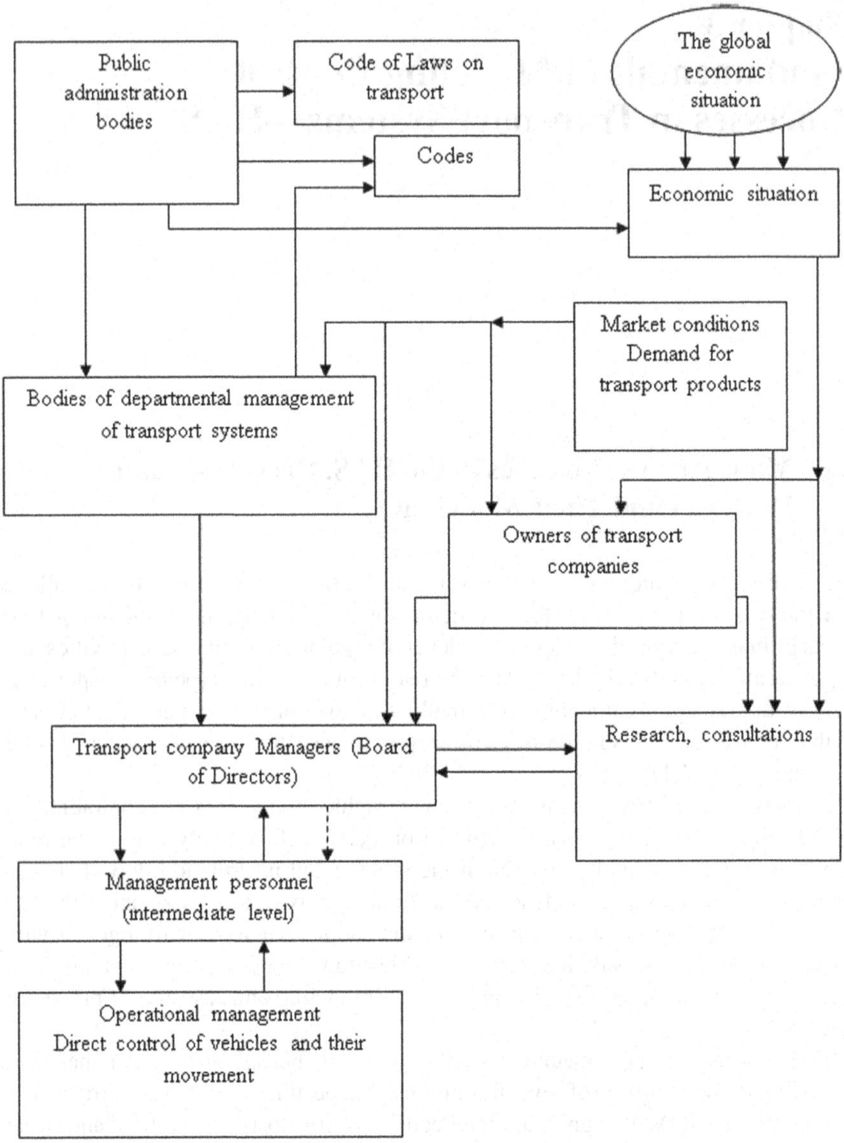

Fig. 3.1 Typical structure of transport systems management

taking into account the limitations that always exist in the functioning of the HAS. It is only with the help of DMP theory methods that it is possible to assess claims and desired goals from a unified position, taking into account available resources and limitations.

It is clear that both operators and researchers of the TSy have been working on the development of such methods and approaches for a very long time. In the last 30–40 years, a scientific direction has been formed—the theory of decision-making processes (DMP), for which the central points are the issues of developing the goal of supporting the DM, finding out the essence of how the DM develops and DMP, i.e. the principle of the solution, and how it can and should be helped (supported) in complex selection tasks. Any new scientific direction is created on the basis of some already known and related directions. The theory of DMP is connected with the study of operations, cybernetics, control theory, the theory of artificial intelligence, while maintaining its own, different from other areas and its inherent tasks and methods of their research.

The main difference between the DMP is the nature of the DMP tasks, when it is necessary to find (choose) and make a decision, the consequences of which will become apparent only in the future—for the operator of the transport process—upon its completion, for the manager—after summing up the economic results of a given period. These consequences of decisions cannot be objectively assessed by mathematical methods before the start of DMP, since such assessments are carried out according to various indicators-criteria and, as a rule, are contradictory: One solution option gives the greatest effect according to one criteria, and the other—according to others, and there is no objective mathematical method for determining the best compromise between these solutions.

However, the transport process cannot be stopped, and decisions must be made. Most often, this volitional act of the DM is carried out on the basis of intuition and experience. Thus, the uncertainty that has arisen in assessing the consequences of the DMP is compensated by the information that the DM is the carrier of. It should be emphasized that this uncertainty is of a fundamental nature and cannot be eliminated by applying an even more complete model of this transport process and a more subtle method of objective assessments.

The DM, which is responsible for its consequences, by its act of choosing a particular decision, gives preference to a criterion or group of criteria of a certain kind (security, guarantee, stability, or, conversely, the greatest speed, profit, risk, etc.). Taking into account these preferences is the most important moment in the DM. The fact is that such preferences serve, on the one hand, as a reflection of the capabilities of a person as a biological species (the human factor (HF)—in terms of speed, the number of moments taken into account, etc.), and also show his (DM) personal attachments of character, habits, degree of training, his level of responsibility, creative handwriting, the operator's line of behavior (personal factor (PF)), etc.—on the other hand.

Thus, the totality of the decisions taken by this DM carries (contains) information, albeit of a subjective nature (taking into account the HF and the PF), which allows the integration of the DM and the HAS into a single system, which ultimately gives the most complete assessment of the options for the decisions taken.

The phrase "subjective nature" reflects the presence of two factors the HF and the PF, as well as the fact that in the HAS—TSy of any kind, the DM must be rational at least in order to be able to justify and explain to others the meaning and essence of its decision. Hence, there is no doubt about the closeness of the space of choice of the DM and its preferences in the DM when managing transport processes in the HAS within some closed rational system—a certain line of behavior, school, policy, often adopted and characteristic of this group of DM with common views. It is this unity of views that largely determines the prestige and overall economic position of the vehicle or its part (transport campaign, station, road, traffic control or their associations, etc.).

In typical situations for the TSy with insufficiently defined consequences of decisions taken by the DM in a dynamically changing environment, so far no cybernetic unit is able to replace the ability of a well-selected, prepared and healthy DM to generate hypotheses, make a forecast, make a decision, bringing the situation out of critical, or, preferably, with their competent DMP and actions to prevent the situation is critical.

Thus, the management of processes in the vehicle is often not possible to determine the "objective" optimal solutions, and the quality of "subjective" solutions is significantly determined by the HF and PF of this DM, as well as methods and procedures for the development and justification of DMP. Hence, it becomes obvious (justified) and the choice that was made to improve the quality of process management in the DMP: It is necessary to compensate for the influence of the HF and PF DM (development of methods and procedures for standard solutions, support for DMP, improvement of the entire system of training DM), as well as to develop a set of methods and procedures for borrowing rational DMP from experienced, successful DM in the spirit of solving inverse optimization problems.

It can be considered that the applied theory of decision-making in the HAS combines research on the psychology of studying and describing real decision-making procedures, on mathematical axiomatic models of choice processes and, finally, on the synthesis of normative methods that recommend (defining, prescribing) DM rules of rational choice. The mathematical theory of choice is currently advanced far enough, and its results will be taken into account one way or another when considering the methods and procedures of DMP in the HAS [1–7], similarly, methods of justifying collective choice, a relatively new branch of collective voting selection processes, will be involved only indirectly.

The main attention here is paid to the development of regulatory methods of DMP in the HAS and the use of the results of solving the inverse problem of optimization of DMP [3–6], as well as the level of operational direct management, which consists in the development and decision-making. Therefore, it can be argued that the main actor in the management of transport systems is the decision-maker (DM), and the main management processes are decision-making processes. At the same time, it is clear that decisions that can be nurtured, maintained and justified for a long time get a completely different color than decisions that should be made immediately—at the current time of movement of vehicles. So there are differences in the DMP and not only in relation to the time of the process itself, but also in relation to the situation that has arisen in which the management DMP is taking place, as well as in the degree of fame of the TSM process itself—can the mathematical model of the TSM process be considered known—or is there none. It would be possible to detail these

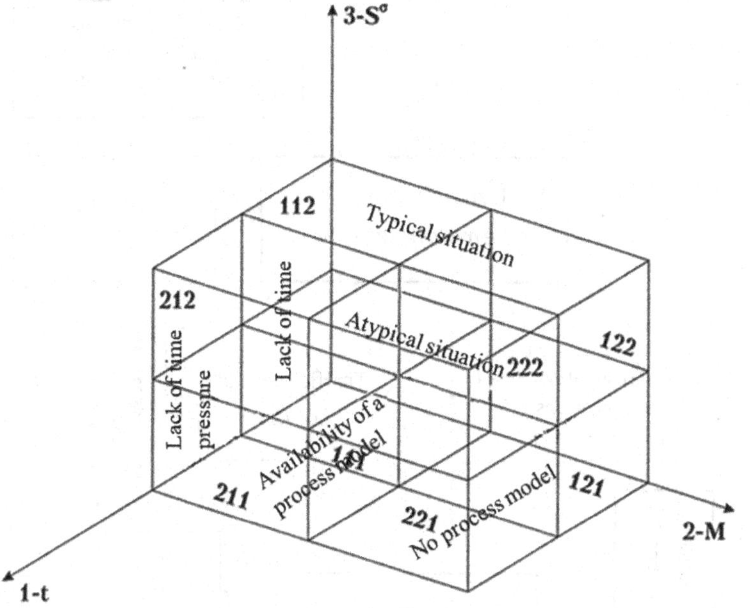

Fig. 3.2 Types of decision-making processes

classification features or take into account some more of the types of DMP that arise during a more detailed analysis of the types of DMP and, consequently, the types of DM activities, but the most significant of them are listed and conditionally divide the space of types of DMP into eight blocks (Fig. 3.2).

And yet, almost 35 years of research experience in the management of DMP allows us to assert that the central factor dividing the DM into two main groups is the presence or absence of a time deficit in their direct DMP.

Indeed, if we return to the typical structure of the TSM (Fig. 3.1), it is clear that each of the blocks can also be deployed in the form of a structural scheme of functioning. So, for example, the operational level of direct control on almost any type of transport allows us to present the following model—a structural control scheme for each of the vehicles in the form of a simple vehicle control circuit (SVCC), regardless of its type (Fig. 3.3).

Thus, it is here that DMP occurs, as a rule, in conditions of time shortage due to the dynamic characteristics of the transport situation (situation), which can be conditionally, quite fully modeled -characterized by a scheme of transition probabilities (Fig. 3.4), where $p_{k-1,k}\left(U^{(\Im)}_{k-1,k}\right)$, $p_{k+1,k}\left(U^{(\Im)}_{k+1,k}\right)$—“inputs” for the situation S_k—the probability of transitions to her from situations S_{k-1} and S_{k+1} by DMP-commands and accepted by the s-th DM; $i^{(\Im)}_{k-1,k}$, $i^{(\Im)}_{k+1,k}$ information flows carrying commands (actions) DM of the s-th level; $p^{(\Im)}_{k,k+1}\left(U^{(\Im)}_{k,k+1}\right)$—"outputs" of the probability that a

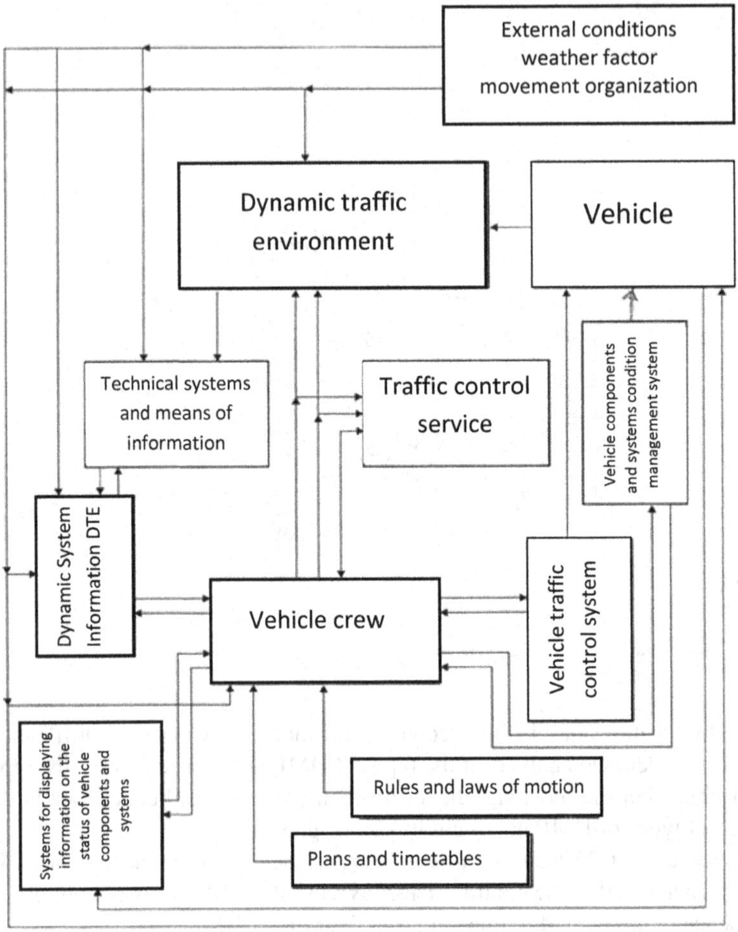

Fig. 3.3 Simplest vehicle control circuit

transition from a given situation S_k will take place to S_{k-1} or S_{k+1} situations by the DMP—command $U^{(\Im)}_{k,k\pm1}$; $i^{(\Im)}_{k,k\pm1}$—corresponding output information flows.

The diagram (Fig. 3.4) shows four characteristic states of the transport situation in the form of transport situations.

S_1—typical (standard)—in the absence of potential conflicts and special cases;

S_2 is complicated when there are potential (seem to be) conflicts and/or special cases, but there are no failures in management;

S_3 is a conflict state of the transport situation when, as a result of the development of one of the potential (apparent) conflicts and/or special cases, management failures occur;

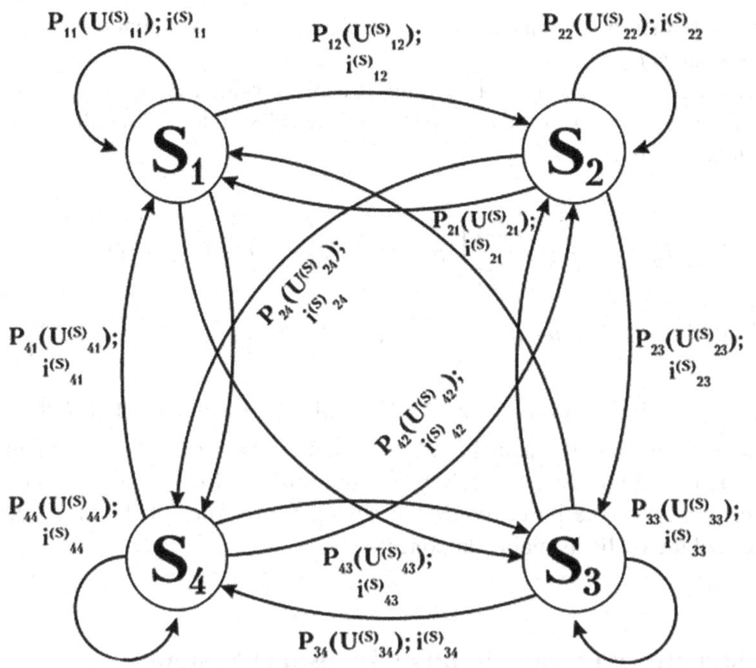

Fig. 3.4 State of the dynamic transport situation

S_4—emergency state of the transport situation—a situation when one of the road users and/or how many in distress with special traffic cases or under special traffic conditions.

Here, control failures are classified as errors in the DMP, leading to a prerequisite for a traffic accident (TA) or incident; and/or a delay in the DMP, which complicate management and also lead to a TSp or incident.

It is clear that such a state of the dynamic transport situation (DTS) is quite conditional. However, it is acceptable—it is sufficient to represent the dependencies of transitions S_1 to $S_{2,3,4}$ and vice versa on the level of professional thinking ability of the DM and its functional management acts, i.e. on the effectiveness of its management activities, which are based on the DMP.

In the HAS, LDPR operators are distinguished, working in conditions of time shortage and also by a greater proportion of physical and motor activity, or intellectual. At the same time, both in the first and in the second cases, the bottleneck in DMP is its information support.

It is for this reason that in the diagram (Fig. 3.4), in addition to probabilities $p^{(\Im)}_{k,k+1}\left(U^{(\Im)}_{k,k+1}\right)$, the values of information flows are also given $i^{(\Im)}_{k,k\pm 1}$.

Thus, the effectiveness of the DMP can be assessed by probabilistic measures of the transitions of the dynamic transport situation (DTS) from one state to another, and therefore thereby obtain an assessment of the final management product, which the teams act as $U_{k,k\pm1}^{(\Im)}$.

Then the total amount of the informing stream, carrying exactly such commands $\Im$ DM $\Im$ that allow you to change the state of the DTS in the desired direction, can be cordoned off by:

$$\Delta\left(i_{(k,k-1)}^{(\Im)}; i_{(k,k)}^{(\Im)}\right) = \frac{h_k^{(\Im)}}{2}\left\{\sum_{r=1}^{h_k^{(\Im)}}\left[\left|p_{(k,k-1)}\left(U_{(k,k-1,r)}^{(\Im)}\right) - p_{(k,k-1)}(O_r)\right| + \left|p_{(k,k)}\left(U_{(k,k,r)}^{(\Im)}\right) - p_{(k,k)}(O_r)\right| - \left|p_{(k,k+1)}\left(U_{(k,k+1,r)}^{(\Im)} - p_{(k,k+1)}(O_r)\right|\right]\right\} \tag{3.1}$$

where $U_{(k,k\pm1,r)}^{(\Im)}$ is the r-th command $\Im$ of the DM in the conditions of the k-th DTS in order to transition it to the state S_{k-1}; $h_k^{(\Im)}$—coefficients, parameters experimentally set for each s-level of the DM, each type of transport, and O_r—the pass command, introduced just to take into account the dynamic properties of the entire control process and the DMP, its main component.

3.2 Structural Organization of Transport Systems Management

Regardless of the type of decision-making processes, their quality, assessed by a certain, reasonable set of performance indicators, is significantly determined and depends on the quality of the structural organization of the HAS. Other things being equal, the activity of the DM at the HAS depends, among other things, on the structural conditions of traffic—at the lower levels of management, on the structural structures of the legal conditions of the activities of transport companies or their aggregates—at higher, regional levels, on the structural organization of legislation, economics and other national institutions of federal significance—at the highest level management of transport systems at the federal level.

Hence, it also becomes clear that the quality of the activities of the DM of each next level can be assessed by the "gain" in the additional effect that is obtained as a result of the activities of the DM of this level: All efforts for the structural organization of the HAS at this level of the hierarchy are undertaken at the next level. If we present the hierarchical typical structure of the HAS (Fig. 3.1.) in a more formalized form, then a well-known hierarchical structure of the fan type is formed (Fig. 3.5). The practice and theory of management systems, including transport systems, convincingly testify to the possibility and legality of supplementing the conditional classification of the DM also in relation to the following additional features: The number of participants in the DM allows you to identify several cases—with one DM ($N_{\Im} = 1$), a team of equal DM ($\text{DM}_1 - \Im \approx \text{DM}_2 - \approx \ldots \approx \text{DM}_{N\Im} - \Im$), one DM $- \Im$ and a group of expert consultants, etc.; the degree of opposition in the HAS highlights DMP

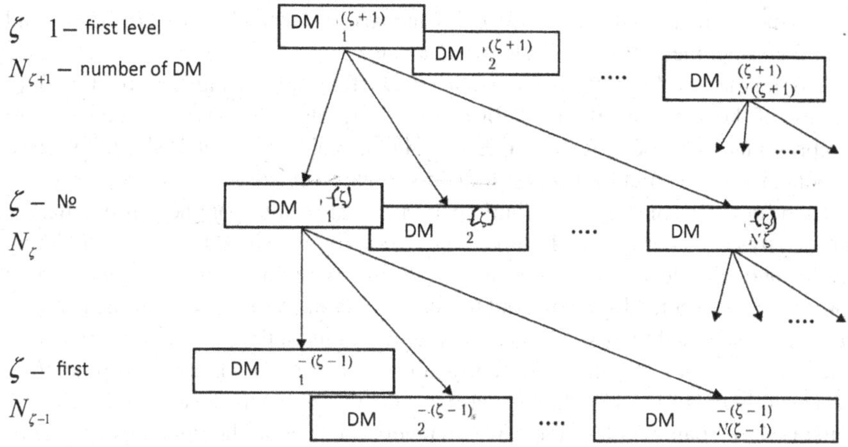

Fig. 3.5 Hierarchical structure of the interaction of the fan-type DM

in conditions of antagonism or consistent interests, etc.; the information support of the DMP, which distinguishes three main options—the option of full awareness of the DM, the presence of interference and fuzzy source information, and, finally, the option of complete uncertainty.

At the same time, it is easy to notice that the number of such signs of the classification of DM can be significantly increased, for example, by introducing a more extended classification of DMP by types of interaction schemes and relationships of DM and others. It is only important to note that almost all types of DMP and classes of LDPR, respectively, are clearly traced when trying to model the activities of the DM at the levels of the hierarchy of the ergatic subsystem of transport systems management. Indeed, if we evaluate the essence of management at each of the levels, then we can note both their general structure and principles of DMP, and differences in their types according to the additional classification given above (Fig. 3.2).

So, for example, at the lowest level of the hierarchy of the vehicle—with direct control of vehicles and their movement (flows or individual units) in the simplest circuit (Fig. 3.3)—DMP is carried out, as a rule, in conditions of time shortage with one DM with consistent interests—a typical situation in the presence of interference and fuzzy initial information, and the object of control (in the case of movement) is a dynamic transport situation (DTS)—a set of vehicles in the environment and zones of operation of technical devices.

There are also cases of DMP with opposite interests, i.e. DMP in a conflict or complicated conflict situation (Fig. 3.4).

Another type of DMP is characteristic of higher levels of the hierarchy—when directing the change of the lower-level DM, when choosing options for the structure of the movement or technical controls, etc.

Here there are many variants of the DSS, taking into account various schemes and mechanisms of the relationship of the LDPR, determined both by the established management practice at the lower levels and by the whole set of governing and regulatory documents, as well as legislative acts.

A fundamental difficulty in the study of DMP and the evaluation of the effectiveness of the activities of the LDPR of various levels of hierarchy in the structure is represented by modeling—complete mathematical models of DMP, taking into account at least the main aspects and factors, do not yet exist.

Therefore, reasonable and reliable experimental results obtained in practice or during physical (semi-natural) modeling of simulator, simulating and modeling devices and complexes can be considered so important and significant [3, 6]. Such studies can most often be carried out to obtain quantitative estimates of the parameters of the DMP and their subsequent mathematical modeling. These are usually the time characteristics of delays in speech-functional acts (SFA), errors in predicting the development of DTS or other situations—depending on the level of the hierarchy of the DM, the bandwidth of the DMP channel according to the time lag of the DM, depending on the level of his so-called professional thinking abilities (PTA)[3, 6].

At the same time, as a rule, it is enough to consider a simplified model of production and DMP on, which of course does not fully reflect the abundance of all kinds of filters and direct and feedback loops when processing information in analyzers and blocks of DM models.

A variant of such a simplified model of the development of DMP in the form of a final speech-functional act (SFA) (mechanical action, speech command-voicing) is presented in (Fig. 3.6) where v = 1,2,… shows multi-alphabetic, i.e. multi-channel perception of the DM information.

According to the concept of single-channel DMP, proposed by academician P. K. Anokhin back in 1946 and has not been refuted to date, the more effective the control effects of the DM are, the less time passes from the moment of receipt of input information to the moment of the SFA, and the more accurate its prognostic activity.

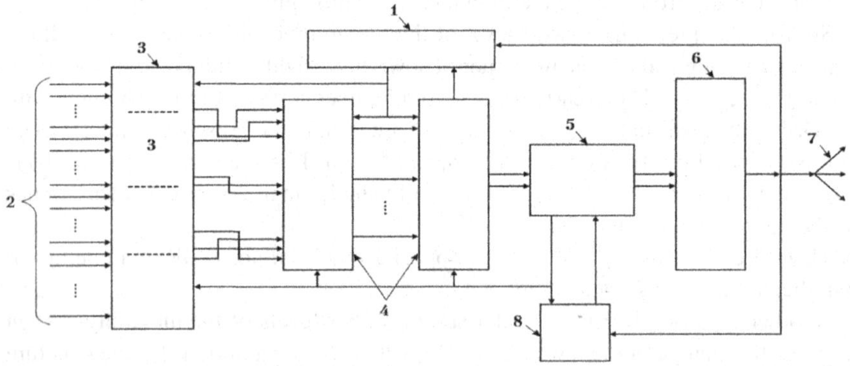

Fig. 3.6 Simplified model of formation SFA. (1) Short-term memory block; (2) Input physical stimuli; (3) Sense organs; (4) Filters; (5) Single-channel decision-making unit; (6) The mechanism of the central effect; (7) Results of SFA—DTS situations; (8) Long-term memory block

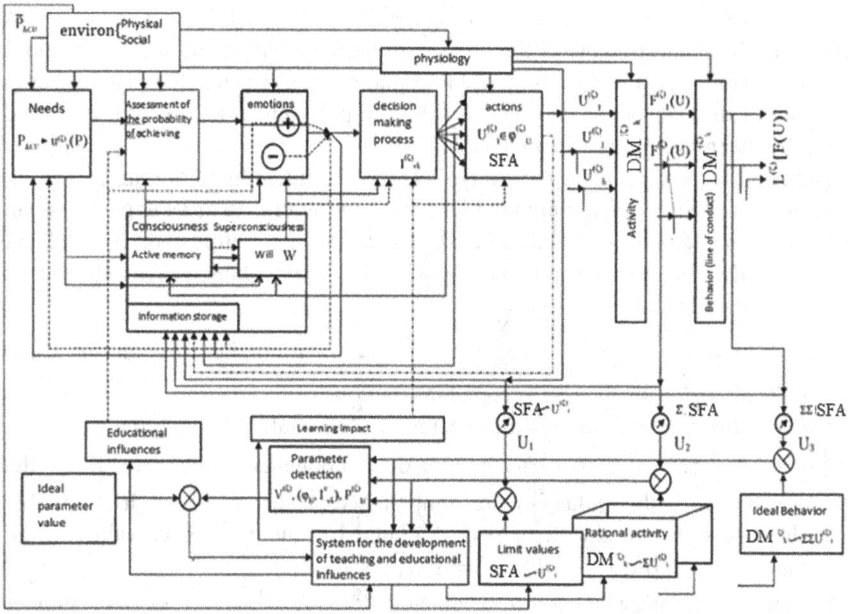

Fig. 3.7 Simplified structure of SFA formation

Taking into account the aspects of professional training of DM and the need to maintain a sufficient level of their PTA in the DMP, it is possible to propose a more detailed simplified structure for the formation of the SFA (Fig. 3.7).

Experimental studies within the framework of such simplified models and structures are quite successful for the practice of HAS and allow us to evaluate the main indicators of the effectiveness of DM in vehicle management. If we start this process "from below", then we can establish the dependence of the values of the performance indicators of the $DM_k(S-1)$, $k = 1, N_{s\text{-}1}$ on the level of traffic organization, technical equipment and legal security, i.e. on the effectiveness of the $DM_k(\Im)$,$k = 1, N_\Im$, $DM_k(\Im + 1)$, $k = 1, N_{\Im+1}$ etc.

If it is possible to link, for example, the circumstance of the course of the DM (DTS, the level of technical equipment, technological regulations of the DM and its psychophysiological state), then we can talk about the concept of DM workload—a vector value, the components of which, for example, in the I-th approximation are its routine employment, determined by the technology of activity, tension, determined by the emerging fatigue, characterized by the general psychophysiological state of the DM and its fitness, is also a situation in the management of the vehicle.

Then, in equal other conditions, the effective activity of the DM can be achieved only by minimizing its workload. It is because of this fact $DM_k - \Im$ that the magnitude of the decrease in the workload vector, etc., can be taken into account as an indicator of activity $DM_k - (\Im - 1)$.

Thus, the DM workload vector can be accepted as one of the indicators of the DM activity at the highest levels of the hierarchy in the development and/or structural decisions. This is all the more important to understand, because the main effect—the profit of the transport company—is achieved precisely at the lower level of the hierarchy when managing the vehicle [1].

Hence, there is a need for the study of DMP and attempts to optimize them by more effective professional training of DM or by automation in order to parry human and/or personal factors, just to take into account the entire structure of the hierarchical system of DM, for example, at least in a simplified form, when at each s-th level of the TSM are allocated:

- the set of solutions $\left\{U_l^{(\Im)} \in L_u^{(\Im)}\right\}, l = \overline{1, m_\Im}$—the number of solutions at the $\Im$-th level; $L_u^{(\Im)}$—the area of acceptable solutions;
- the set of components of the vector of the state of the control object $\left\{x_i^{(\Im)} \in L_x^{(\Im)}\right\}, l = \overline{1, n_\Im}$—the number of components of the state; $L_x^{(\Im)}$—the range of permissible values of the components; at the same time, it is possible to distinguish (highlight) the expected, planned, actually obtained values of the components $x_i^{(\Im)}(t)$; $y_i^{(\Im)} \in L_y^{(\Im)}; i = \overline{1, n_{y\Im}} \le n_\Im$;
- the set of situations when driving a vehicle $\left\{S_r^{(\Im)} \in L_s^{(\Im)}\right\}, r = \overline{1, k_\Im}$—the number of situations taken into account at the s-th level; $L_s^{(\Im)}$—the area of probable situations at the s-th level;
- the set of probabilities of choosing the given k-th DM_k – $\Im$-th solution: $\left\{p_{\mathrm{kl}}^{(\Im)} \in L_p^{(\Im)}\right\}$;
- a set of components of the state vector of the control object measured or represented and displayed by someone or something:

$$\left\{Z_i^{(\Im)} \in L_Z^{(\Im)}; Z_i^{(\Im)}(t) = \sum\nolimits_{j=1}^{n_{\Im Z}} a_{ij}\left(x_j^{(\Im)}(t) + ya\right) + \xi_i^{(\Im)}(t); i = \overline{1, n_{\Im S}}, \overline{\overline{n_{ZS}}} \le \overline{\overline{n_\Im}}\right\}, r = \overline{1, k_\Im}$$

$\xi_i^{(\Im)}(t)$—a random function or quantity that characterizes the measurement errors of the components of the state vector or its distortions during presentation, $Y\left(y^\Im - x_i\right)$—a characteristic of the deviation of the state from the planned;

- a set of private performance indicators, $\left\{I_{\nu_{\Im k}}^{(\Im)} \in J_Z^{(\Im)}\right\}, \nu_{\Im k} = \overline{1, \mu_{\Im k}}$—the number of private indicators at the $\Im$-th level; among the set, a subset of indicators reflecting individual values for the k-th $\mathrm{DM_k}$ of any j-th result achieved in the DMP process at the $\Im$-th level should be distinguished:

$$\left\{I_{\rho k}^{(\Im)} \in L_\Sigma^{(\Im)}; \rho_k^{(\Im)} = \Im + f^k, x \le \mu_\Im, I_\rho^{(\Im)} = I_\rho^{(\Im)}\left[\bar{\bar{x}}_i^{(\Im)}(t), \bar{U}_l^{(\Im)}\left(S_r^{(\Im)}, t\right)\right] \in J_\Sigma^{(\Im)}\right\}$$

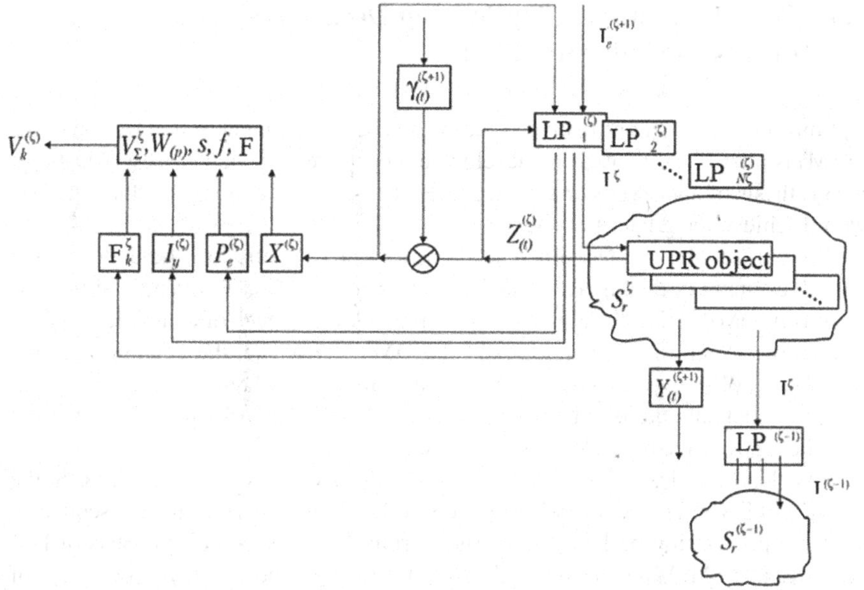

Fig 3.8 Block diagram of the model of the functionally generalized characteristic of the personal factor of the DM

The structure of the interaction of the DM at the levels of the hierarchical HAS system in terms of the introduced parameters of the DM and such characteristics of the DM as its preferences $I_\rho^{(\Im)}$ and awareness $\left\{Z_i^{(\Im)}(t)\right\}$ of the state of the control object can now be represented as (Fig. 3.8).

Thus, the effectiveness of the DM-($\Im + 1$) can now be assessed by the degree of success of the structural organization of the working conditions for the DM-$\Im$, whose activity can be defined in terms of the parameters of the DM and the vector of its workload. It should be noted that the main characteristics of the DMP parameters introduced during modeling are amenable to experimental studies within the framework of a real HAS or on the basis of training and simulating devices or their complexes. Based on individual actions in the storage devices of such training complexes, and often in practice, if they have sufficient automation tools, some generalizing functions $F_k^{(\Im)}\left[U_l^{(\Im)}(I, x, S)\right], L_k^{(\Im)}\left[F_k^{(\Im)}(I_\nu, x(t), S, U)\right]$, can be formed that characterize the activity and behavior of this k-th DM when managing processes at this level in this structure and situation S.

3.3 The Main Tasks of Optimizing Decision-Making Processes Under the TSM

Optimization of decision-making processes in the management of transport systems (TSM) is possible only after the development of models of such processes involves the synthesis of algorithms or at least some rules, acting according to which the DM would achieve the greatest efficiency.

The phrase "the greatest efficiency" has a clear semantic load here, describing the achievement of the highest value by the indicator of the effectiveness of the DM activity in DMP. All of the above seems to imply the presence of complete certainty in the description of the conditions of the DM activity (i.e. the conditions of the functioning of the vehicle at this level of its management hierarchy), a fairly clear explicit understanding of what can be called "good" management and what is the best, i.e. knowledge of quality assessments.

It is clear that this is not the case in all the diverse cases that often occur in the practice of TSM. Hence, the whole variety of DMP models and, as a consequence, a lot of optimization problems that arise during the analysis and synthesis of DM activities at TSM. When trying to classify or compile a more or less ordered list of such tasks, it is possible to use the classification of DMP and, accordingly, types of DM in TSM (Fig. 3.2), as well as models in the form of structural schemes of the hierarchical system of TSM (Figs. 3.1, 3.5, 3.8). As a result of compiling such an ordered list, it is possible to obtain a certain set of tasks for optimizing the DMP for the DM of each type encountered in practice TSM and for the main situations that arise in this case (Fig. 3.4).

At the same time, at a time when almost all federal and regional TSy carry out their activities in market conditions, the problem of constructing indicators of the effectiveness of DMP in developing systems is of great interest. Indeed, any vehicle can be attributed to such systems: A developing system is called a system that has not only system-wide principles of decomposition and aggregation of processes, but also their increase in complexity, number of connections, tasks, hierarchy levels, etc. The greatest interest is precisely the construction of reasonable indicators of the effectiveness of DMP with TSM in a developing type of system.

At the same time, for any TSy, a state vector is defined at the $\Im$-th level of the hierarchy $X^{(\Im)} = \left(x_1^{(\Im)}, \ldots, x_{\mathrm{nk}}^{(\Im)}\right)'$, a vector of control functions $U_k^{(\Im)}(S_p) = \left(U_{1k}^{(\Im)}, \ldots, U_{\mathrm{mk}}^{(\Im)}\right)$ in the s-th situation $\left(\rho = \overline{1,4}\right)$ for who DM-$\Im\left(k = \overline{1,M}, S = \overline{S_1} - \overline{S_4}\right)$- control parameters $P_k^{(\Im)}(S_p) = \left(P_{1k}^{(\Im)}, \ldots, P_{\mathrm{rk}}^{(\Im)}\right)$.

The DMP efficiency indicator for TSM in the form of:

$$J_k^{(\Im)} = F\left[J_{\Sigma}^{(1)\Im}, \ldots, U_{\Sigma}^{(M)\Im}\right]$$

$$J_{\Sigma}^{(\Im)k} = F_k^{(\Im)}\left[I_1\left(U_k^{(\Im)}, P_k^{(\Im)}, f_k^{(\Im)}\right), \ldots, I_{\mu k}^{(\Im)}\left(U_k^{(\Im)}, P_k^{(\Im)}, f_k^{(\Im)}\right)\right],$$

$\left(\Im = \overline{1,M}, k = \overline{1,N_{\Im}}, \nu = \overline{1,\mu}\right)$; $f_k^{(\Im)}$—unknown accidental exposure.

It is clear that the designations are introduced for each specific process of the developing system. As a key point here, the case of the presence of the fact that the flow of the process is influenced by not the same $DM^{(k)}$, but several: $DM_k^{(\Im)} k = \overline{1, N_{\Im}}$.

Thus, the optimization of PPR is given here under conditions of special uncertainty due to the presence of several $DM_k^{(S)}$ interests (aspirations, motivations, etc.) that do not necessarily coincide. The essence of the problem lies in the formalization of a common performance indicator for all $DM_k^{(S)}$ in the form of a convolution. It is necessary to first introduce an assumption about the type of compromise between the DM. Such assumptions can be reduced to the conditions for the emergence of a team-collective, when the group $DM_k^{(S)}$ is considered as a single whole, combining its own $U_k^{(\Im)}$ and $P_k^{(\Im)}$ achieving optimization of the DMP according to a single indicator found using an iterative procedure: It assumes the discipline of each $DM_k^{(S)}$ positive motivation regarding the effectiveness of the DMP in the developing vehicle under consideration, as well as their almost uniform interpretation of regulatory legal acts-documents, it can be considered a condition for the formation of a single team-a team of DM in the form of justification and construction of a single performance indicator.

The development of the system through the emergence or expansion of the hierarchy is one of the most common cases in the practice of TSy activities. Then there is a typical, for example, fan structure (Fig. 3.5), when the group $DM_k^{(\Im)}$ $\left(k = \overline{1, N_{\Im}}, \Im = \overline{1, M}\right)$ at the $\Im$ level of the management hierarchy, in turn, obeys the decisions made by the head of the DM-k who has now been allocated to the $k+1$ level, DM-k or $DM^{(k+1)}$ often happens when the group $DM_k^{(\Im)}$ managing the processes of transport production in the TSy works under the leadership of (management) of one DM—commander or senior person of the DM-k, or, given the designation of the hierarchy level,—$DM^{(k+1)}$. At the same time, the most significant point here is the fact that the interests of the team $DM_k^{(\Im)}$ directly managing the production process and the interests of the commander of this team no longer completely coincide.

This means that the DM team and the DM-k have their own, non-matching performance indicators.

From the point of view of each of the k-th $\left(k = \overline{1, N_{\Im}}\right)$ team members is:

$$J_k^{(\Im)}\left(U_k^{(\Im)}, P_k^{(\Im)}, f, U^{(\Im+1)}, P^{(\Im+1)}\right) \tag{3.2}$$

as well as the indicator they form that is common to the entire team:

$$J_k^{(\Im)} = F^{(\Im)}\left[J_k^{(\Im)}\left(U_k^{(\Im)}, P_k^{(\Im)}, f, U^{(\Im+1)}, P^{(\Im+1)}\right)\right] \tag{3.3}$$

and an indicator of the efficiency of the PR process at the TCB in terms of:$DM^{(\Im+1)}$:

$$J_k^{(\Im+1)}\left[U_k^{(\Im+1)}, P_k^{(\Im+1)}, f, U_k^{(\Im)}, P_k^{(\Im)}, J_k^{(\Im)}, J_{\Sigma}^{(\Im)}\right] \tag{3.4}$$

The essence of the task is that the DM-k, i.e. $DM^{(\Im+1)}$ is interested in the results of the flow of production processes, which it manages only indirectly through the collective $DM_k^{(\Im)}$, this explains the presence in its indicator $J_k^{(\Im+1)}$ as management functions $U_k^{(\Im)}$, $P_k^{(\Im)}$, the implementation of which is carried out $DM_k^{(\Im)}$. From the DM-k side, control is carried out using commands $U_k^{(\Im+1)}$ and parameters $P_k^{(\Im+1)}$. The influence of random influences is indicated by the presence of a random argument f. Among all the possible variety of possible options for interaction between DM-k and $DM_k^{(\Im)}$ the most interesting is the option when DM-k participates only indirectly in the indicator of the flow of the production process and, in addition, it gives the "installation" first, informing the team $DM_k^{(S)}$ about its DMP, i.e. about its controls $U_k^{(\Im+1)}$ and parameters $P_k^{(\Im+1)}$.

This is in good agreement with the conduct of debriefings on air transport, the conduct of flyovers or selector debriefings on road and rail transport. It is also logical to assume that DM-k has full information regarding DMP each of the $DM_k^{(\Im)}$, that is, he knows $U_k^{(\Im)}$, $P_k^{(\Im)}\left(\Im = \overline{1, M}\right)$.

This almost always corresponds to the awareness of an experienced DM-k regarding the DMP of each of the team members: This happens, for example, in the duty shift of air traffic controllers. When the flight manager (DM-k) knows almost at every moment what is happening in each of the $DM_k^{(\Im)}$ management sectors, which is led by one of the dispatchers $U_k^{(\Im)}$, $P_k^{(\Im)}\left(\Im = \overline{1, M}\right)$.

At the same time, it is logical to assume that each of $DM_k^{(\Im)}$ them forms an indicator comparable to that of another member of the team. This is the division of the overall load, the choice of one of the tasks from the set of tasks, etc. This can be represented in the form of dependencies showing that any part of the value of the k-th $DM_k^{(\Im)}$ performance indicator can be expressed as a proportional part of the indicator of another i-th $DM_i^{(\Im)}$ from the team:

$$\Delta_{\mathrm{ki}}^{(\Im)} = \Delta J_{\sum k}^{(\Im)} = h_{\mathrm{ki}} \cdot \Delta J_{\sum i}^{(\Im)}$$

with:

$$\Delta_{\mathrm{ik}} \cdot \Delta_{\mathrm{kj}} = \Delta_{\mathrm{ij}};\ \Delta_{\mathrm{ki}} \cdot \Delta_{\mathrm{ik}} = 1 \tag{3.5}$$

Now for the simplest types of reconciliation of indicators F in (3.3), for example, such as:

$$J_{\sum}^{(\Im)} = \min k\left[\omega_k\left(J_{\sum k}^{(\Im)} - J_{\min k}^{(\Im)}\right)\right],\ \left(k = \overline{1, N_\Im}, \Im = \overline{1, M} = 4\right) \tag{3.6}$$

The general indicator (3.2) can be represented by the expression:

$$J_{\sum k}^{(\Im)} = J_{\sum \mathrm{Ok}}^{(\Im)} - \sum\nolimits_{i=1}^{N_\Im} \Delta_{\mathrm{ki}} + \sum\nolimits_{i=1}^{N_\Im} \omega_{\mathrm{ik}} \Delta_{\mathrm{ik}} \tag{3.7}$$

where $J^{(\Im)}_{\sum \mathrm{Ok}}$—the initial, before the introduction of compensating increments, its value.

It is clear that all the variety of relationships in the team between $\mathrm{DM}_k^{(\Im)}$ and between the DM-k and the members of the team is far from being exhausted by the incremental compensation scheme. Taking into account the uncertainties that arise when trying to account for such relationships between $\mathrm{DM}_k^{(\Im)}$ the team, on the one hand, and between the team $\mathrm{DM}_k^{(\Im)}$ and the DM-k, on the other, causes great difficulties. Considering that the vehicle can always be attributed to high-risk systems for which approaches related to life support, and therefore with the applicability of the guarantee approach, when the calculation is for the worst case. With this approach, each of $\mathrm{DM}_k^{(\Im)}$ them acts (carries out the DMP by choice $U_k^{(\Im)}$, $P_k^{(\Im)}$) thus, in order to proceed from his maximum strategy, i.e. so that he maximizes the efficiency indicator in conditions when the latter is already minimized by the management and parameters of another team member. This can be represented by the expression:

$$J^{(\Im)}_{\min}\left(\tilde{U}_k, \tilde{P}_k\right) = \max_{\substack{U_k \in \phi_U^{(\Im)} \\ P_k \in \phi_P^{(\Im)}}} \min_{\substack{U_i \in \phi_U^{(i)} \\ P_i \in \phi_P^{(i)}}} \left[J^{(\Im)}_{\sum k}\left(U_k^{(\Im)}, P_k^{(\Im)}, U_i^{(\Im)}, P_i^{(\Im)}\right)\right] \tag{3.8}$$

$\left(i = \overline{1, k}, \overline{k, N_\Im}, r\right), \phi_U, \phi_P$—the areas of acceptable values of controls and parameters for each of all at all $\mathrm{DM}_k^{(\Im)}$ levels of the hierarchy.

The most important property of such a guarantee strategy of DMP in the form $\tilde{U}_k, \tilde{P}_k$ of any of them $\mathrm{DM}_k^{(\Im)}$ is that each of them can achieve the value by independent actions, without taking into account the actions of other DM, including DM-k, therefore, when forming a team of DM with their overall performance indicator in the latter using convolution (3.6) as the smallest value values not less than This condition should be taken into account, along with the conditions (3.5), serve as conditions for the stability (expediency) of the DM team, since guarantees a significant increase in the effectiveness of DMP in the case of collective actions of DM_k than in the case of their individual actions:

$$\max_{U_\Sigma^{(\Im)} \in \phi_U^{(\Im)}} \left\{ J^{(\Im)} = \min_k \left[\omega_k \left(J^{(\Im)}_{\sum k} - J^{(\Im)}_{\min k} \right) \right] \right\}, \left(k = \overline{1, N_\Im}\right) \tag{3.9}$$

where $U_\Sigma^{(\Im)} = \left(U_1^{(\Im)}, \ldots, U_{N_\Im}^{(\Im)}\right)'$—the management function of the entire DM team, considered as a whole.

This form of performance indicator (3.9), considered as an assessment of the activities of the entire team, allows for sufficient flexibility to reflect all possible compromise relations between the DM_k.

Indeed, if we take two extreme cases for the general presence of comparability of the indicators of each of the DM, i.e. the presence of compensating increments in both Δ_{ik} (3.5) and Δ_{ki} (3.6), then in the first of them let:

$$\sum\nolimits_{k=1}^{N_{\Im}} \omega_{\text{ki}}^{(\Im)} \Delta_{\text{ki}}^{(\Im)} - \sum\nolimits_{k=1}^{N_{\Im}} \Delta_{\text{ik}}^{(\Im)} = 0; \sum\nolimits_{j=N_{\Im}+1}^{N_{\Im}} \omega_{\text{ji}}^{(\Im)} \Delta_{\text{ji}}^{(\Im)} = 0,$$

where $j = \overline{N_{\Im}+1, N}$—indexes denoting DM_j that are not part of the team-collective. This immediately means from (3.9) that each of the indicators $J_{\sum k}^{(\Im)}$ will be taken from (3.7) without compensating increments, i.e.:

$$J_{\sum i}^{(\Im)} = J_{\sum \text{Oi}}^{(\Im)}; J^{(\Im)} = \overset{\min}{k}\left[\omega_k\left(J_{\sum O\Im}^{(\Im)} - J_{\min k}^{(\Im)}\right)\right], (k = \overline{1, N_{\Im}})$$

Thus, the evaluation of the indicator from the point of view of the DM team is equivalent here to the evaluation of a multi-criteria process, and the compromise between the DM in the team is completely equivalent to the compromise between the indicators in a multi-criteria problem.

If now, for example, one of the ways to choose coefficients ω_k, then the uncertainty caused by the difficulty of formalizing the relationship between the DM in the team can be considered eliminated. If you still take in the second case

$$\sum_{k=1}^{N_{\Im}} \omega_{\text{ki}}^{(\Im)} \Delta_{\text{ki}}^{(\Im)} - \sum_{k=1}^{N_{\Im}} \Delta_{\text{ik}}^{(\Im)} = \tilde{\Delta}_i; \text{ but at the same time } \sum_{i=1}^{N} \tilde{\Delta}_i = 0, \tag{3.10}$$

then the DM team, in order to optimize the production process, can, in addition to control functions $U_{\sum}^{(\Im)}$ and parametric controls $P_{\sum}^{(\Im)}$, also choose values $\tilde{\Delta}_i^{(\Im)}$—a new type of control parameters that form a vector for the DM team $\tilde{\Delta}^{(\Im)} = \left(\tilde{\Delta}_1^{(\Im)}, \ldots, \tilde{\Delta}_{N_{\Im}}^{(\Im)}\right)'$.

The choice $\tilde{\Delta}_i^{(\Im)}$ is then made on the terms:

$$\overset{\max}{\tilde{\Delta}^{(\Im)}}\left\{\overset{\min}{k}\left[\omega_k^{(\Im)}\left(J_{\sum \text{Ok}}^{(\Im)} + \tilde{\Delta}_k^{(\Im)} - J_{\min k}^{(\Im)}\right)\right]\right\} \tag{3.11}$$

which leads, as is known, to the necessary optimality condition in the form of equality of weighted indicators for all $DM_k^{(\Im)}$ [3]:

$$\omega_i^{(\Im)}\left[J_{\Sigma 0i}^{(\Im)}\left(U_i^{(\Im)}, P_i^{(\Im)}, f\right) + \tilde{\Delta}_i^{*(\Im)} - J_{\min i}^{(\Im)}\right] = R^{(\Im)}, i = 1, \overline{N}_{\Im} \tag{3.12}$$

From this condition $\omega_k^{(\Im)} > 0$ it follows that:

$$J_{\Sigma Oi}^{(\Im)} + \tilde{\Delta}_i^{*(\Im)} = J_{\min i}^{(\Im)} + \frac{R^{(\Im)}}{\omega_i^{(\Im)}}; R^{(\Im)} \sum\nolimits_{i=1}^{N_{\Im}} \frac{1}{\omega_i^{(\Im)}} = \sum\nolimits_{i=1}^{M} J_{\Sigma i}^{(\Im)} - \sum\nolimits_{i=1}^{N_{\Im}} J_{\min i}^{(\Im)}$$

where from

$$\begin{aligned} J_{\Sigma k}^{(\Im)} &= J_{\Sigma \mathrm{O}k}^{(\Im)} + \tilde{\Delta}_k^{*(\Im)} \\ &= J_{\min k}^{(\Im)} - \frac{1}{\omega_k^{(\Im)} \sum_{i=1}^{N} \frac{1}{\omega_i^{(k)}}} \left(\sum\nolimits_{i=1}^{N} J_{\Sigma i}^{(\Im)} - \sum\nolimits_{i=1}^{N_\Im} J_{\min i}^{(\Im)} \right) \end{aligned} \tag{3.13}$$

The conditions obtained for the case (3.10) of the optimality of parametric control of the collective of the type $\tilde{\Delta}_i^{(\Im)}\left(i = \overline{1, N_\Im}\right)$ indicate that the vector $\tilde{\Delta}^{*(\Im)} = \left(\tilde{\Delta}_1^{*(\Im)}, \ldots, \tilde{\Delta}_{N_\Im}^{*(\Im)}\right)'$, when the condition (3.12) is fulfilled, determines the degree of influence of each individual's $\mathrm{DM}_k^{(\Im)}$ on the value of the overall indicator of the efficiency of the production process:

$$J^{(\Im)} = \sum\nolimits_{\Im=1}^{N_\Im} \omega_k^{(\Im)} J_{\Sigma k}^{(\Im)} \tag{3.14}$$

This also confirms once again the equivalence of the two types of convolution of performance indicators for the team (3.6) and (3.14) in the case of the optimal vector $\tilde{\Delta}^{*(\Im)}$.

So, the guarantee approach used for the organization of the team allows you to specify the conditions of its expediency from an objective point of view DM-k in the form of a condition fulfillment requirement:

$$\begin{gathered} J_{\min k}^{(\Im)} \geq \left.\begin{matrix} U_\Im^{(k)} \\ \overline{P_\Im^{(k)}} \end{matrix}\right\}^{\max} \in \varphi_{\mathrm{UPk}}^{(\Im)} U^{(i)} J_{\Sigma k}^{(\Im)} \\ \omega_i^{(\Im)}\left(J_{\Sigma O}^{(i)} + \Delta_i^{*(\Im)} - J_{\min i}^{(k)}\right) = R^{(\Im)} > 0 \end{gathered} \tag{3.15}$$

along with the conditions (3.5).

The activity of the team in these conditions is aimed at increasing the value of the overall process efficiency indicator (3.9) and choosing such $\Delta^{*(\Im)}$ that the conditions of its stability and expediency are met (3.11–3.15).

Optimization of the DMP at the TSM can be carried out depending on the conditions of activity n $\mathrm{DM}^{(\Im)}$ at this level of the hierarchy of the management system.

The main features of such conditions of activity are:

- the presence or absence of a mathematical model of controlled processes in the vehicle (reasonable performance indicators), dependencies that determine the patterns of changes in the vector of the state of the production process in the vehicle, restrictions that determine the range of permissible values of control functions and parameters, and sometimes coordinates;
- the presence or absence of a shortage of time for DMP—the development of control values and parameters;
- the presence or absence of information (measurements) or its incompleteness, necessary for the development of controls and parameters, i.e. the presence or absence of uncertainty conditions in TSM.

In addition, when formulating the entire possible set of tasks for optimizing the DMP in TSM, there are practically no formulations (statements) of tasks where the task would be to determine the degree of effectiveness of the DMP in TSM, depending on:

(1) the duration of the analysis of the state vector and the time of the DMP;
(2) the completeness and type (!) of the presentation of the DM information for the implementation of the DMP;
(3) the depth and accuracy of the forecast of the possible development of the area of the vector of the state of the production transport process in the vehicle;
(4) the values of the elements of speech interaction, as well as the time of conducting speech interaction between DM of different levels; and finally,
(5) the degree of professional training of the DM, the characteristics of its human and personal factors (HF, PF) at the DMP.

The practice of carrying out the activities of the DM under the TS, as well as the projects of transport systems management systems, shows that the optimization of the DM can be considered from different positions. Of greatest interest is the automation of the DMP, i.e. automated search and presentation of the DM support in the DMP, optimal from the point of view of a reasonable indicator. It is equally important to find a number of "home" optimal DMP blanks for the case where the automation of the TSy is not carried out due to the lack of technical means (computers, information display media, communications, etc.). Finally, the optimization of the processes of preparation and training of DM for their implementation of DMP at TSM, close to optimal, should also be attributed to the most important, main tasks of optimizing DMP at TSM.

One of the possible options for the formulation of the generalized optimization problem of DMP in TSM can be represented as follows:

Let a given $\Im$ level of the hierarchy of the vehicle and the conditions of the organization, legal acts in the form of regulations of activity, vehicle traffic rules, TSy rules TSM, etc., traffic plans or other restrictions are known. The generalized task is to implement such a DMP by choosing such a set of controls and parameters:

$$U_k^{(\Im)} = \left(U_{k1}^{(\Im)}, \ldots, U_{\mathrm{km}}^{(\Im)}\right);$$
$$p_k^{(\Im)} = \left(p_{k1}^{(\Im)}, \ldots, p_{\mathrm{kQ}}^{(\Im)}\right);\ \left(k = 1, N_{\Im};\ \Im = 1, \bar{M}\right) \tag{3.16}$$

out of the range of acceptable values:

$$U_k^{(\Im)} \in \varphi_U^{(\Im)},\ \left(k = 1, \overline{N}_{\Im}\right);\ p_q^{(\Im)} \in \varphi_p^{(\Im)},\ \left(q = 1, \overline{Q}_{\Im}\right) \tag{3.17}$$

at which the state vector of the controlled production transport process:

$$X^{(\Im)}(t) = \left[x_1^{(\Im)}(t), \ldots, x_{n_{\Im}}^{(\Im)}(t)\right] \tag{3.18}$$

known $DM^{(k)}$ incompletely and inaccurately or with sufficient completeness and accuracy using measurements or intuitively, based on experience, i.e. $DM^{(k)}$ has knowledge about the state of the production transport process in the TSy at the level of knowledge (representation) of some other vector

$$Z_k^{\Im}(t) = \varphi\left[X^{(\Im)}(t), f_k^{(\Im)}\right]; Z_k^{(\Im)}(t) = \left[Z_{k1}^{(\Im)}(t), \ldots, Z_{k\alpha}^{(\Im)}(t)\right] \tag{3.19}$$

where ϕ is the functional dependence connecting the components of the vectors $Z_k^{(\Im)}(t)$ and $X^{(\Im)}(t)$ under the conditions of interference or uncertainty conditions set by a parameter or function $f_k^{(\Im)}$, as well as the ratio of the number of their components n and $\Im\alpha$.

To implement the DMP, the DM makes a forecast of the development of the situation in the TSy, i.e. determines the assessment

$$Z_k^{(\Im)}(t+\tau) \to X'^{(\Im)}(t+\tau) = \varphi^{-1}\left[Z_k^{(\Im)}(t+\tau), f^k\right] \tag{3.20}$$

During the period $t + \tau$, it will change from the initial state to the final one (at this stage) according to a known (or not specified) pattern of the form:

$$L\left[X^{(\Im)}(t), U_k^{(\Im)}, p_k^{(\Im)}, X^{(\Im)}(t+\tau)\right]_{\leq}^{\geq}[] \tag{3.21}$$

so that the efficiency indicator of such a transition and similar management reaches the optimal value:

$$\begin{aligned} J_\Sigma^{(\Im)} = F^{(\Im)}\Big[& I_1^{(\Im)}\left({}^{\backprime}X^{(\Im)}, U_k^{(\Im)}, p_k^{(\Im)}, t, t+\tau\right), \ldots, \\ & I_\mu^{(\Im)}\left({}^{\backprime}X^{(\Im)}, U_k^{(\Im)}, p_k^{(\Im)}, t, t+\tau\right)\Big] \end{aligned} \tag{3.22}$$

where $I_1^{(\Im)}, \ldots, I_\mu^{(\Im)}$—a set of particular performance indicators, the type of which is known or understood only at a qualitative level and requires justification and explication in the form of functional dependencies of the particular for each of the $I_\nu, \left(\nu = \overline{1, \mu}\right)$ and/or general type $F^{(\Im)}[I()]$.

Now practically each of the above-described tasks of optimization of DMP in TSM can be formulated in terms of a generalized problem (3.16–3.22). For example, often in the practice of TSM there are, as they say, comprehensively successful solutions, and the efficiency indicator is not functionally defined—not found. Such a problem is called the inverse optimization problem, and for this case, it is the inverse optimization problem of the DMP at TSM—and is formulated as follows.

Find the type of performance indicator (3.22) and the permissible set of its parameters if a set of successful (optimal) controls and parameters (3.16) is given for the transition of the process (3.18) from the initial state to the final one according to the regularity (3.21) or according to the forecast $DM_k^{(\Im)}$ (3.20).

The formulation of the task of optimizing the learning process (preparation) is of great difficulty DM for the implementation of optimal DMP. Here the concept of the components of the state vector is shifted. Now, as such a vector, it becomes not a vector of the state of the production process, but a vector of the degree (level) of DM training. If such a replacement is carried out and the functional dependencies of the change of such a vector are known under the influence of the controls of the effects of the educator (instructor, teacher, trainer), then the task of optimizing the learning process can also be formulated in a similar way.

Among the optimization tasks of the DMP TSM, the tasks of estimating the costs of the DM, which he does at the DMP in the TSM, are of particular importance. Such tasks are interesting not only from the point of view of labor rationing at various $DM^{(\Im)}$ at the $\Im$ levels of the TSM hierarchy, they are of considerable interest and from the conditions of the flow of production processes in the TSy, as well as during the selection of candidates for the preparation of DM for TSM, choose those of them whose costs (psychophysiological, volitional, etc. parametrically measured) are the smallest or do not exceed the set values. Here, as before, substitutions can be used in terms of the generalized optimization problem of the DMP during TSM.

3.4 Methods of Research and Optimization of DMP in TSM

The study of the DM in general, and especially in the case of the TS, involves the creation of formalized models or at least a structural representation of the activities of the DM at various levels of hierarchy for their explication in the future. At all such levels, the creation of formalized models of professional activity of the LDPR is still a fundamental scientific problem in the CU.

The complexity of the processes, starting with the presence of random outliers and nonlinear properties of control objects and the presence of ever-increasing requirements for control efficiency, introduces its own specifics into the methods of modeling and research of DMP at TSM. The main specific feature here is the almost complete coverage of all types of mathematical models when trying to model DMP in TSM: from the simplest logic circuits to the most complex intelligent models used for control. When modeling the DM, it is also necessary to take into account the limitations that are inevitably introduced by the presence of such an element of systems as the DM at all levels of the hierarchy. Here it is important to emphasize once again that there is a certain limit to human capabilities in DMP—the human factor (HF), just as there is a certain handwriting in the activities of each DM—the personal factor (PF), the modeling of which is still far from its completion. The principal difficulties in modeling the DMP in the TSM are: taking into account uncertainty factors in the creation and presentation of process models in management systems at different levels of its hierarchy, taking into account the presence of different performance indicators for different DM or the presence of different preferences.

In all known methods of research of the DMP, such as methods of mathematical programming optimization of dynamic processes of various kinds, group solutions, as well as using game theory, one of the main directions of the efforts undertaken was just such difficulties. The methods related to or based on methods of mathematical programming and optimization of dynamic systems have found the greatest application, as well as development in research, modeling and optimization. Widely used methods of multi-criteria optimization are mainly reduced to the establishment of additional conditions under which the problem that has arisen in one way or another is reduced to a scalar performance indicator or to a sequence of tasks with one indicator.

The DMP procedures obtained in this way allow them to be found quite often and actively used in TSM. In group solutions, procedures for matching aggregated (composite) solutions with individual preferences are effective. It is often possible to develop a procedure that is convenient for DMP with the help of compromises and so-called fair decisions and conditions within the framework of game theory. All these directions fit into the general structure of the knowledge base and database for their application in the general theory of TSM. At the end of the 80s and throughout the 90s of the twentieth century, the number of works on the creation of intelligent control systems, in general, and the use of artificial intelligence methods in the TSM, including, increased dramatically. At the same time, considerable funds are being spent on developments in this area: In Japan alone, about $6 billion was spent on the development of intelligent control systems in 2000, the USA planned 3 billion for these purposes, and the countries of Western Europe in total—more than 7 billion. Such intelligent-looking TSM are already being used at the present time, for example, when controlling space flights in the USA, controlling the movement of electric trains of the metro line in Naboku in Japan, when creating flight control systems for aircraft in Russia, etc.

The intelligent systems used in the TSM are systems based on interdisciplinary developments, information technologies and informatization of the DMP at the TSM. The concept of artificial intelligence arose from the need to combine various scientific knowledge into a single scientific picture of the world. The main difference between intelligent types of TSM is the presence in them of special blocks-structures that carry out directed, controlled accumulation of information, its processing and development on this basis of DMP-tips (supports) for DM, or, finally, with full automation of DMP - their implementation with TSM. The use of intelligent control systems as part of hierarchical control systems of transport systems, for example, at the lower level of the hierarchy, where transportation is carried out, makes it possible to increase the overall efficiency of the TSM due to the emerging opportunity to use more data in motion. At the same time, it becomes possible to perform such functions as:

- automated planning of traffic routes in order to increase the capacity of the transport space, traffic safety, minimize fuel consumption, increase the coordination of traffic in the DTS;
- automated route planning on individual traffic sections with bypass of emergency or congested areas of space to overcome obstacles or weather events;
- automated assessment of hazards based on data from all sources of information and evaluation of the effectiveness of various methods of their parrying;

- automatic control of the modes of operation of all means of receiving and processing information in order to ensure its centralized use for DMP;
- possible coordinated use of information from the systems of a certain group of vehicles to increase the total amount of available information for DMP in the control system of each vehicle, etc.

The implementation of these tasks finds its embodiment most often in expert systems and neural networks. At the same time, neural networks are most often used in control systems for the operation of individual nodes and vehicle systems for their diagnostics based on solving such complex and complex tasks as detecting equipment failures. Neural networks are used in the form of independent control systems, or integrated with traditional systems.

Expert systems are used for the development and DMP at the TSM of the widest purpose—as an advising and supporting DMP of the DM of any level of the hierarchy at the TSM. The general structure of the inclusion of intelligent systems in one of the levels of hierarchical systems for the development of a functional act in the TSM is shown in (Fig. 3.9).

It should be emphasized that research in the field of intelligent control systems has allowed us to formulate and justify the principle of IPDI (Increasing Precision with Decreasing Intelligence), which states that the higher the control accuracy, the lower the intelligence of the system behavior. This principle makes it possible to determine the degree of intelligence of the support units of the DMP at various levels of the hierarchy in the TSM: The higher the level, the higher the degree of intelligence should be, the wider the knowledge base—to develop a common management strategy (in terms of fuzzy sets, for example, using linguistic SFAs). At the lowest levels—the most accurate tips and support in the DMP do not require a high degree of intelligence of the system with direct control. In cases when an intelligent system is connected with TSM instead of DM with full automation of control processes, it is called active and, if it is an expert-type system, then such an active expert system should contain an output SFA converter of exactly the type that corresponds to the inputs of traditional control systems of the hierarchy level where it is used.

From the above brief overview of the methods and systems of DMP at TSM, it follows that there is already a well-known conclusion about the existence of such a scientific direction as decision-making. The specific features that such a field of application gives it, such as transport in general and management of transport systems in particular, have already been noted and allow us to outline the main directions for the application of research methods and optimization of DMP, which in the greatest way allow to increase the efficiency of TSM:

- compensating and/or parrying the influence of PF and HF by developing DMP methods and procedures to support LDPR;
- automation of DMP processes at TSM, including intelligent systems for the same purposes or replacement of LDPR;
- improvement of the DM training system with the use of DMP optimization methods;

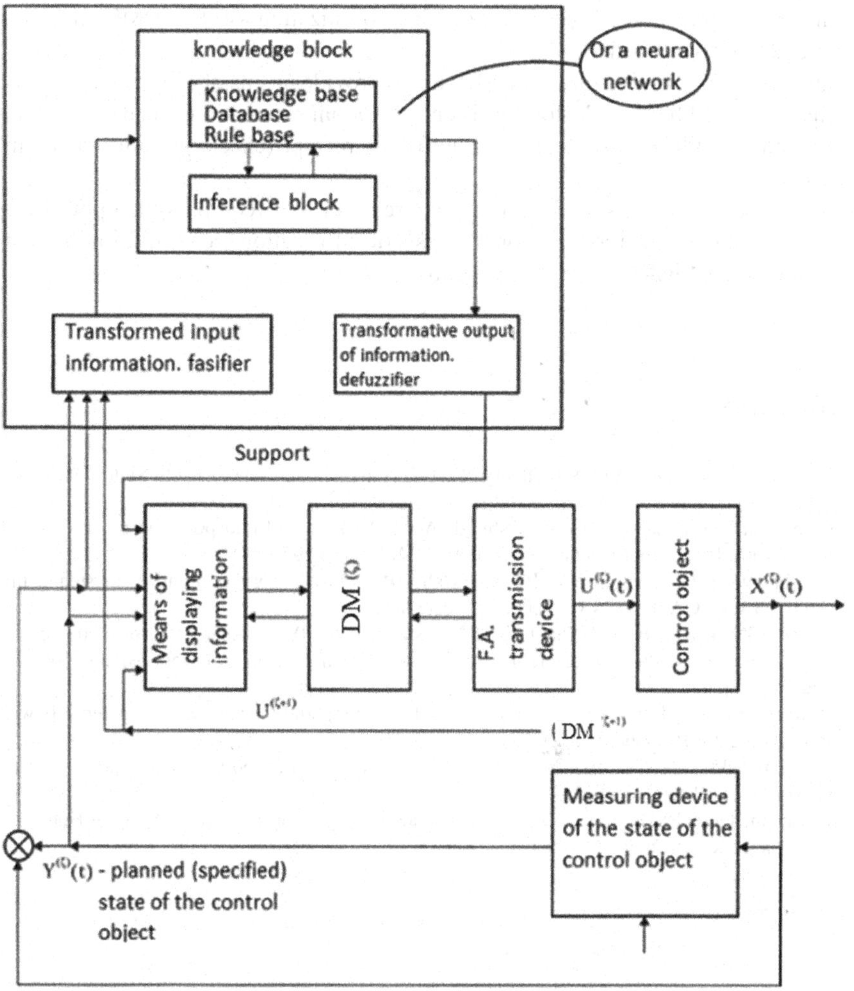

Fig. 3.9 General structure of the inclusion of an intellectual system (IS) in the system of speech production-a functional act in the TSM

- development of methods and procedures for borrowing rational DMP from experienced and successful DM when they carry out their activities under the TSM.

Then we can also name a number of requirements for the used or developed methods of optimization of DMP:

- any methods and methods of obtaining from the LDPR or providing (displaying) information processing should correspond to the capabilities of a trained (trained) human operator or manager for the perception and processing of information;

- any assumptions introduced during the optimization of the DMP must be mathematically justified;
- any methods of optimizing the DMP should provide means of checking the information issued by the DM for consistency (random reservations, random outliers in actions with management, etc.) during its receipt (checking with maximum speed);
- at high levels of the OTS hierarchy, any relations between possible options for DMP should be justified only on the basis of information received from the DM of this or the highest level of the hierarchy.

References

1. Burkov VN, Novikov DA (1999) Theory of active systems: state and prospects. Sinteg, Moscow, p 128
2. Galaburda VG, Persianov VA, Timoshinidr AA (1996) Unified transport system: studies for universities. In: Galaburdy VG (ed). Transport, Moscow, p 295
3. Kryzhanovsky GA, Kupin VV, Plyasovskikh AP (2008) Theory of transport systems. In: Kryzhanovsky GA (ed) GA University, St. Petersburg
4. Zaitsev EN, Bogdanov EV, Shaidurov IG, Pesterev EV (2008) General course of transport: a textbook for the study of discipline and the performance of control work. SPbGUGA, St. Petersburg, p 98
5. Palagin YI (2009) Logistics. In: Palagin YI (ed) Planning and management of material flows. Polytechnic, St. Petersburg, p 286
6. Kryzhanovsky GA, Shashkin VV (2001) Management of transport systems Part 3. Severnazvezda, , St Petersburg, p 224
7. Krasnoshchekov PS, Petrov AA (1983) Principles of model construction. Publishing House of Moscow State University, Moscow, p 264

Chapter 4
Information, Modeling and Measurement of Uncertainty in the Vehicle

4.1 Information Support of Management and Decision-Making Processes in the HAS

The activity of any transport system, as well as the activity of any HAS, is determined by the quality of its information support (IS). It is clear that the activity of the vehicle depends on the effectiveness of management in it, i.e. on the effectiveness of the content and accuracy of the representation of direct (control) and reverse (control) links. The number of communication channels is determined by the elements $a_{i\mu}$ and a_{ij} the matrix with C. But the amount of information and its quantity determines the decomposition, i.e. the structure of the S system. Information flows through direct and feedback links, as well as through channels of interchange between system elements and with elements of the external environment, determine the entire volume of information in the vehicle. When this volume increases to the values that are limiting for the perception of the DM, an automated information system—the AIS of the vehicle is needed in any vehicle.

Indeed, comparatively simple reasoning supports this statement. Suppose, for example, a specific vehicle company has a goal (and there are enough orders for this) aimed at a sharp increase in the number of vehicles used. The activities of each of them are managed by one of the company's representative offices in this region. A certain number of flights generates a certain flow of information, and with an increase in the number of vehicles, such a flow increases. At the same time, each representative office should act in conjunction with related representative offices in this region, as well as in other regions. The problem arises about the rational number of such representations.

Its solution would be obviously simple if it were possible to construct curves of dependence of the values of the information flow on the number of representative offices and the number of vehicles. It is obvious that with the increase in the number of representative offices, the information flow, determined by the number of vehicles, decreases: $\frac{\partial I^{(1)}(N,n)}{\partial N} < 0$, and the information flow of approvals, on the contrary, is

© The Author(s), under exclusive license to Springer Nature Singapore Pte Ltd. 2023
G. A. Kryzhanovsky et al., *Modeling of Transportation Aviation Processes*, Springer Aerospace Technology, https://doi.org/10.1007/978-981-19-7607-0_4

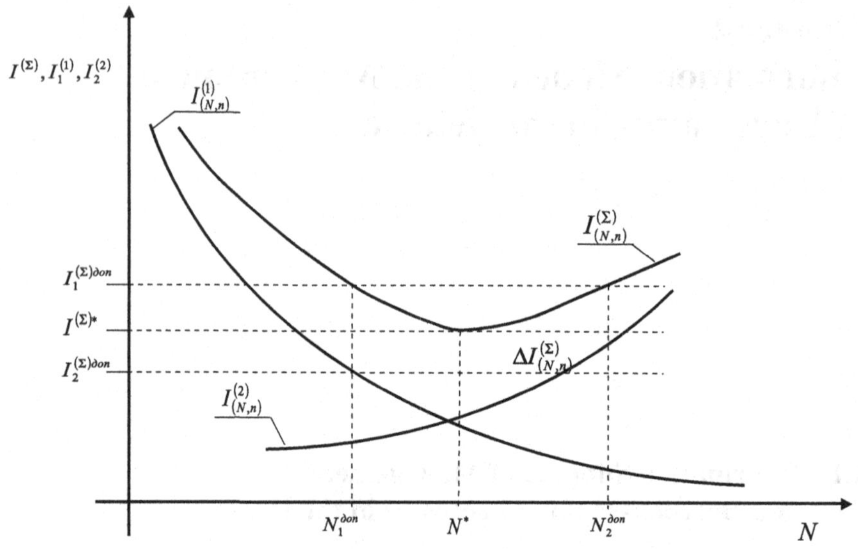

Fig. 4.1 Decomposition of the vehicle. Information flows in the vehicle

growing: $\frac{\partial I^{(2)}(N,n)}{\partial N} > 0$. If you enter the value of the total flow $I^{(\Sigma)} = I^{(1)}(N,n) + I^{(2)}(N,n)$, and its acceptable value for DM $I^{(\Sigma)\partial on}$, then the rational number of N^* representations will be obtained from the ratio $\frac{\partial I^{(\Sigma)}(N,n)}{\partial N} = 0$ (see Fig. 4.1). Curves $I^{(1)}(N,n)$ и $I^{(2)}(N,n)$ can be found experimentally, which will allow us to justify the decomposition of the system into a rational number of representations.

Case No. 1: $I_1^{(\Sigma)\partial on}(N,n) \geq I^{(\Sigma)*}(N,n)$: $N_1^{(\partial on)} \leq N^*$.

Case No. 2: $I_2^{(\Sigma)\partial on}(N,n) \leq I^{(\Sigma)*}(N,n)$: $N^* \leq N_2^{(\partial on)}$.

Case No. 3: $I_3^{(\Sigma)\partial on}(N,n) \approx I^{(\Sigma)*}(N,n)$: $N(N,n) \approx N^*(N,n)$.

Case No. 2—proof of the need to implement AIS TSy to compensate for excessive processing $\Delta I^{(\Sigma)}(N,n)$.

Any AIS vehicle includes a number of basic types of software: mathematical, software, linguistic and technical. Mathematical support is formed from mathematical models, methods and algorithms of information processing and its aggregation to represent the DM in the form most convenient for perception (awareness). The software is a set of application programs, operational documentation for them, which, along with technical equipment, are necessary for processing and presenting information.

Technical equipment includes all computing devices, organizational equipment and means of transmitting and receiving data that interact with each other and form the AIS of the vehicle.

The following groups of technical equipment can be distinguished:

- groups of data preparation and input necessary for automating the processes of entering alphanumeric and iconographic information into a computer, encoding, compression and visualization;
- a group of information exchange, remote communication of technical means of various levels of hierarchy and elements of the AIS of the TSy through various communication channels;
- a group of computers for receiving digital data from data input tools, software processing, accumulation and output of information to machine media (drives), to information transmission channels or its display;
- a group of displaying information about decisions made, in the form of displays, printers, plotters (plotters) and external storage devices;
- a group of information storage devices in the form of reference data, archival materials, information processing tools for storage, restoration, reproduction and simplification of access.

A significant role in the information support is played by a set of language tools for the formalization of commands in the SFA and other data contained in the forward and backward links of the vehicle. Often, such a set of linguistic means forms a special, formalized from natural language, the language of subject (air, sea, etc.) communication. For example, the phraseology of radio exchange in aviation. Sometimes this also includes the language of document flow, a set of rules for communication and information exchange, as well as communication of the DM with the AIS computer, etc.

In large transport companies, information support is based on AIS, which are based on automated data banks and knowledge banks consisting of databases and knowledge, management systems, presentation and protection of information. When forming such banks, the main principle is the principle of information unity, according to which only a single system of designations, terms, symbols, methods of aggregation and presentation of information, a single dimension of physical quantities, a special language of exchange between the elements of the vehicle is used.

As has been repeatedly noted, information flows in the TSy, like any AIS, contain, along with useful information, a significant part of false, excessive, distorted information, which is generally referred to here by such a concept as uncertainty.

The most striking example in this regard is the aggregate information about the state of the system and the alleged actions of the opposing side when forming an image of the situation in the DM.

4.2 Modeling and Measurement of Uncertainty. A Priori and a Posteriori Entropy, Measurement of Uncertainty in DMP

To model various kinds of uncertainties that have to be taken into account in management (i.e. in the analysis, development of DMP and control actions), the development of measures to prevent transport incidents, when assessing the degree of reliability of any of the events in the chain of development of confrontation, the general state of the system or the dynamic transport situation, the concept of probability is usually used.

However, classical probability theory is limited by the scope of application in which the law of large numbers is valid and the axiomatics of A. N. Kolmogorov is valid. A significant, if not overwhelming part of the events in the DTS are rarely recurring events, and for their modeling the basic hypotheses of classical probability theory are not feasible.

In recent decades (since the 60s of the last century), various concepts have been used in technology to model (describe) uncertainty or reliability. Here are some of them: vague sets, intuitive probability, subjective probability, random, psychological, logical, empirical probabilities, relative frequency, degree of conviction and degree of rational opinion. At the same time, each of these concepts has its own axiomatics and a limited scope of application.

The concept of risk, assessed on the basis of such representations of uncertainty or reliability, as well as the definitions themselves, fits into these limited areas—and no more.

The basis here is the general property of DM behavior models, which consists in continuous assessment and prediction of the reliability of events, creating an image of the situation and its development taking into account the state of the environment, including with direct control and management of DTS.

This inevitably leads to attempts to build a model of thinking, which, as we know, is still far from completion. It is practically impossible to deduce consistent constructions of proofs of the truth of the judgments of the DM in conditions of insufficient knowledge about the nature of their judgment, which indicates a dead end outcome of such an approach.

Under these conditions, abandoning attempts to construct estimates of the probability of rarely recurring events, it is possible to offer estimates obtained by processing expert schemes, checking them for consistency and logical rigor. Since each expert is a subject, it is necessary to begin with the introduction of the concept of subjective probability.

Subjective probability of an event K_S the essence is the degree of expectation by the expert S of the chances of occurrence of this event compared to the chance of the opposite event $\overline{\mathrm{K}}_S$, depending on the totality of knowledge, experience and individual characteristics of the expert S:

$$0 \leq P(\mathrm{K}_S,\ \mathrm{X}_S) \leq 1,$$

where X_S—the totality of knowledge, experience and individual characteristics of the expert.

The concept of entropy was first introduced by Rudolf Clausius in thermodynamics in 1865 to determine the measure of irreversible energy dissipation, the measure of deviation of a real process from an ideal one.

It is clear what the definition is X_S allows a wide interpretation and possibilities for the use of experimental data, including probabilistic knowledge based on classical approaches using the frequency of occurrence of individual situations, phenomena and actions. Hence, the statement about the coincidence of essentially subjective probability, which assesses the chances of the occurrence or non-occurrence of the event K_S, with a vague set evaluating the degree of belonging to one or another of their classes. However, the information approach proposed by K. Shannon is of the greatest interest for measuring uncertainty.

The central point in this approach is the introduced K.Shannon's concept of entropy as a measure of uncertainty:

$$H = -R\sum_{k=1}^{r_i} P_{k_i} \cdot \log P_{k_i}, \tag{4.1}$$

where R—non-negative dimension coefficient; $P_{k_i}\left(k_i = \overline{1, r_i}\right)$—for example, a subjective probability determined by processing the results of an expert survey; an event K_S, it consists in its occurrence or non-occurrence, namely in the th section of the confidence interval of the i-th component, k is the ordinal number in this interval $\left(k_i = \overline{1, r_i}\right)$.

In other words, each amount of information about any parameter required for the DMP can be represented by a certain minimum interval of its values containing the true value of this parameter. It is clear that without additional information (information), such a true value of the parameter required for the DMP can be at any point of such a minimum interval with equal probability, i.e.:

$$\sup H\left(P_{k_i}\right) = -\sum_{k=1}^{r_i} P_{k_i} \log P_{k_i} = -\log P_{k_i},$$

where $r_i = \frac{L_i}{\Delta l_i}$—the number of plots; the value in Δl_i—the permissible accuracy of measuring the true value of the parameter; L_i—the minimum interval covering its true value:

$$P_{k_i} = \frac{\Delta l_{k_i}}{L_{k_i}}, \quad \sum_{k_i}^{r_i} P_{k_i} = 1,; \quad \left(k_i = \overline{1, r_i}\right)$$

Note that L_{k_i}—a non-increasing function of the amount of information about this parameter and the receipt (appearance, presence) of any information about the parameter (its values) only narrows—reduces the minimum interval of its values:

$$L_{k_i} = \phi\left(\mathrm{I}_{k_i}\right).$$

That is, when implementing, for example, AIS and obtaining additional information $\Delta \mathrm{I}^{(\Sigma)}(N,\ n)$ (see Fig. 4.1), the value of the minimum evaluation interval will decrease

$$L_{k_i}^{(1)} \geq L_{k_i}^{(0)} - \Delta L_{k_i}; \Delta L_{k_i} = \phi\left[\Delta \mathrm{I}^{\Sigma}(N, n)\right]$$

which, in the case of, for example, the decomposition problem, will narrow the area—the interval for finding the optimal number of vehicle representative offices in the region (see Fig. 4.1).

If we use this approach, formalizing and choosing (justifying) intervals and probabilities, then all possible uncertainties according to (4.1) can be obtained a formula describing from the independence of the initial uncertainties the total entropy of the forecast of the results of the actions of the opposing companies in the system *S* and DMP:

$$H(P_S) = H(S) + H(Y_\Sigma) + H(D_S), \tag{4.2}$$

where $H(D_S)$—entropy of predicting the consequences of actions, i.e. selected from possible *s* alternatives; $H(y_\Sigma)$—the entropy of choosing a set of particular performance indicators *I* and the method of their convolution, which determine the quality of DMP from a group of such aggregates and methods:

$$H(Y_\Sigma) = -\sum_{\upsilon=1}^{\mu} P(I_\upsilon) \cdot \log P(I_\upsilon) - \sum_{\rho=1} P_\rho \cdot \log P_\rho, \tag{4.3}$$

where μ—the number of all possible partial performance indicators $I_\upsilon(\upsilon = \mu)$; $P(I_\upsilon)$—degree of affiliation I_υ to this discrete set of indicators; P_ρ—the degree to which the convolution ρ method belongs to a vague set of such methods that reduce the selected set of indicators to a single scalar performance indicator y_Σ.

Finally, $H(S)$—entropy assumptions about the state of the system as a whole:

$$H(S) = H(C) + H(P) + H(R) + H(W), \tag{4.4}$$

where *H*(*P*), *H*(*R*), *H*(*C*)—entropies characterizing (measuring) uncertainties associated with the assessment of the state of communications, information support, the location of elements of opposing companies and the state of the external environment.

So, the entropy modeling (measuring) the uncertainty of DMP (4.2) is a priori, i.e. it is a priori entropy. Its value determines the amount of information that is required for the implementation of the DMP.

The values of a posteriori entropy can also be found from expression (4.2), but with different values of probabilities and degrees of membership, since some types of uncertainties will change significantly.

It should only be emphasized once again that the assessment of subjective probabilities and degrees of belonging to some vague sets used in the measurement of uncertainties in the form of entropy are clearly subjective everywhere, and hence, the DMP in AIS can be called objective only to the extent that it is possible to achieve objectivity in the estimates used. It is clear that the greatest effect here can be obtained with the appropriate organization of information support, up to the creation of AIS.

If AIS is used rationally, i.e. to achieve optimization, then there is a problem of rational organization of the management of observations and activities of the transport system at all levels of its functioning. Taking into account the random nature of disturbances and interference in each of the processes at each of the levels, we can use entropy of the form (4.1), which allows us to represent any uncertainty in the form

$$H(I_i) = -\int_{-\infty}^{+\infty} P(I_i) \ln P(I_i) dI_i, \tag{4.5}$$

or under conditions of a priori uncertainty in the definition of interference and disturbances, introducing an interval for the density of the distribution of values I_i in the form of $[-\Delta l_i,\ \Delta l_i]$ with a uniform density distribution law; it is possible to obtain: $P(I_i) = \frac{1}{2\Delta l_i}$.

Then from (4.5):

$$H(I_i) = \int_{-\Delta l_i}^{+\Delta l_i} \frac{1}{2\Delta l_i} \ln \frac{1}{2\Delta l_i} \mathrm{d}I_i = \ln 2\Delta l_i. \tag{4.6}$$

If I_i—information about the *i*-th parameter of the TS*y* process, the law of distribution of values of which is unknown (poorly defined), then the interval Δl_i from (4.6) is defined as:

$$\Delta^{(i)} = \frac{1}{2} e^{H(I_i)} = 0,5 \cdot \exp[H(I_i)] \tag{4.7}$$

Then, for any distribution law of the *i*-th parameter of the process TS*y*, the value of the interval $\Delta^{(i)}$ can be defined from (4.7). For example, a frequently used fuzzy or fuzzy set is represented by a law close to the triangular distribution (Simpson). Then $P(I_i) = \frac{1}{2} e^{H(I_i)}$ and

$$\Delta_T^{(i)} = \sigma_i \left[\frac{3}{2} e\right]^{\frac{1}{2}} = 2,02 \cdot \sigma_i \tag{4.8}$$

For the case of a normal distribution:

$$P(I_i) = \frac{1}{\sigma\sqrt{2\pi}} \cdot e^{-\frac{\overset{\circ}{I}_i^2}{2\sigma_i^2}}$$

where $\overset{\circ}{I}_i$—centered random variable I_i.

Given (4.7), one can also find $\Delta_H^{(i)}$ for the case of a normal distribution law:

$$\Delta_H^{(i)} = \tfrac{1}{2} e^{H(I_i)} = \sigma_i \frac{\sqrt{2\pi e}}{2} \approx 2,066 \cdot \sigma_i \tag{4.9}$$

Thus, the interval $\Delta^{(i)}$ is a generalized in this (entropy) sense indicator of the uncertainty of deviations of the *i*-th parameter of the TS*y* process (HAS). This approach was used to characterize deviations and errors in processes of various physical nature.

The value of the generalized indicator $\Delta^{(i)}$ it turns out to be one of the relatively simple characteristics of the "blurriness" of the values of the i-th parameter of any process. Indeed, generalizing its values of various distribution laws $P(I_i)$—(4.8), (4.9) and others, you can write:

$$\Delta_\Sigma^{(i)} = K_\Sigma^{(i)} \cdot \sigma_i, \tag{4.10}$$

where $K_\Sigma^{(i)}$—(entropy) coefficient Σ the law of the distribution of deviations of the i-th parameter, and σ_i—the value of its standard deviation. It is clear that the values K_Σ have limitations

$$0 \le K_\Sigma \le 2,066.$$

The dependence (4.10) characterizes the uncertainty of the characteristics of the dynamics of deviations, replacing the phase plane "deviations—the rate of deviations" with some "entropy" plane $(K_\Sigma,\ \sigma)$ (see Fig. 4.2).

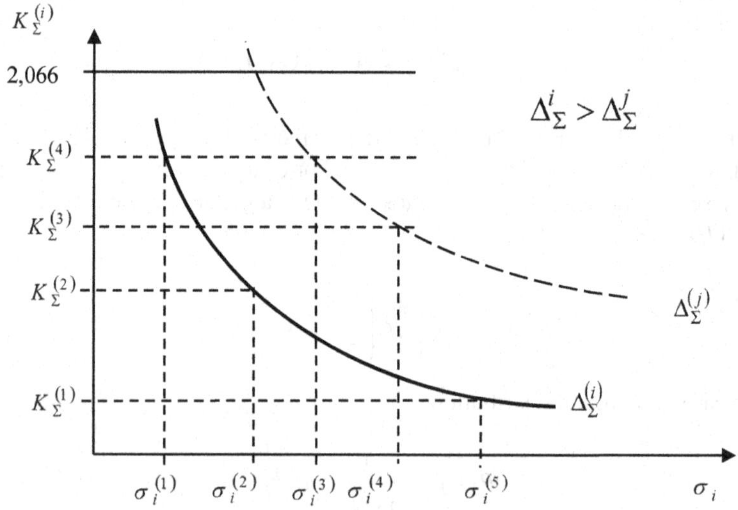

Fig. 4.2 Uncertainty levels determined using the isentropic Δ_Σ line function.

Thus, the level of uncertainty of the values of the i-th parameter can now be set using the functions $f\left(\Delta_{\Sigma}^{(i)}\right) = f\left(K_{\Sigma}^{(i)},\ \sigma^{(i)}\right)$, and if the information (information) about the deviations of the i-th parameter is more accurate than about the deviations of the j-th parameter, then $\Delta_{\Sigma}^{i} > \Delta_{\Sigma}^{j}$.

This means that the solution of the problem of organizing processes in the vehicle and/or managing them should be solved in conditions where

$$\Delta_{\Sigma}^{(i)*} = \min\left\{\Delta_{\Sigma}^{(i)}\right\}, \tag{4.11}$$

that is, with the lowest value of the uncertainty indicator in the values of the i-th parameter.

If we now take as the i-th parameter some output parameter of the control system, then the requirement (4.11) characterizes the requirement of the greatest accuracy of its (i.e. parameter) observation. In this case, an optimization problem arises, for example, in the form of a requirement to determine the best values for controlling the i-th process so that $\sigma^{i}\left(u^{(i)}\right)$ and $K_{\Sigma}^{(i)}\left(u^{(i)}\right)$ would be able to provide the requirements (4.11).

Here is the value of management $u^{(i)}$ determined from the conditions

$$\left.\begin{array}{l} \dfrac{\partial \Delta_{\Sigma}^{(i)}\left(u_k^{(i)}\right)}{\partial u_k^{(i)}} = 0; \qquad \left(k = \overline{1,\ m}\right) \\ \text{and} \\ d\Delta_{\Sigma}^{(i)2}\left(u_1^{(i)},\ \ldots,\ u_m^{(i)}\right) = \sum_{k}^{m}\sum_{l}^{m} \dfrac{\partial^2 \Delta_{\Sigma}^{(i)}}{\partial u_k \cdot \partial u_l} > 0 \end{array}\right\} \tag{4.12}$$

where from

$$\begin{aligned} \partial \Delta_{\Sigma}^{(i)} &= \left[K_{\Sigma}^{(i)}\left(u_k^{(i)}\right) + \partial K_{\Sigma}^{(i)}\left(u_k^{(i)}\right)\right] \cdot \left[\sigma_i\left(u_k^{(i)}\right) + \partial \sigma_i\left(u_k^{(i)}\right)\right] \\ &= K_{\Sigma}^{(i)} \cdot \sigma_i = K_{\Sigma}^{(i)}(u) \cdot \partial \sigma(u) \\ &+ \sigma_i(u) \cdot \partial K_{\Sigma}^{(i)}(u) + \varepsilon(0). \end{aligned}$$

The practical significance of the above approach is that in order to calculate the Δ_{Σ}, a much smaller number of observations is required than when calculating correlation functions using probabilistic and other approaches. The amount of measurement information required for the organization of observations using the approach represented by expressions (4.10–4.12) is determined by the sample size required to calculate the estimates of the COEX—σ_i, since the coefficient estimate $K_{\Sigma}^{(i)}$ is usually carried out empirically on the basis of the physical essence of the i-th parameter of the process occurring in the vehicle. If the i-th parameter is observed in such a way that some process changes are allowed under the influence of control $u^{(i)}$, determined from the conditions (4.12), then it is possible to construct a known gradient process, such as, for example,

$$\Delta_{\Sigma}^{(i)}\left[u_k^{(i)}(h+1)\right] = \underset{u}{grad}\ \Delta_{\Sigma}^{(i)}\left[u_k^{(i)}(h)\right] \cdot \ell + \Delta_{\Sigma}^{(i)}\left[u_k^{(i)}(h)\right], \tag{4.13}$$

where ℓ—step size ($\ell < 0$), h—step number, and since $K_{\Sigma}^{(i)} \approx K_{\Sigma}^{(i)(0)}$, that

$$\underset{u}{grad}\ \Delta_{\Sigma}^{(i)}\left[u_k^{(i)}(h)\right] \cong K_{\Sigma}^{(i)(0)}\left[u_k^{(i)}(h)\right] \cdot \delta\sigma_i\left[u^{(i)}(h)\right].$$

Expression (4.13) is used for a step-by-step process in which

$$\Delta_{\Sigma}^{(i)}(h+1) < \Delta_{\Sigma}^{(i)}(h) < \Delta_{\Sigma}^{(i)}(h-1) < \ldots$$

and so on until the conditions (4.12) are met with an accuracy of up to a small value

$$\varepsilon(0) + \sigma_i\left(h^*\right) \cdot \delta K_{\Sigma}\left(h^*\right) \tag{4.14}$$

Usually, in practice, a priori estimates of values are known $K_{\Sigma}^{(i)}$ and $\sigma^{(i)}$, as well as the limits of their variation for any parameter of the vehicle processes. Then, for each case relevant to a particular situation, the procedure (4.12–4.13) can be used to obtain values close to optimal (up to 4.14) $\Delta_{\Sigma}^{(i)*}$, allowing to determine the value of the i-th parameter with the least uncertainty.

4.3 The Inverse Problem of Optimization of DMP in HAS

In the presence of sufficiently computationally and intellectually powerful AIS in the TSy, there are opportunities to accumulate a significant amount of data and information about the nature of the actions of opposing transport companies, as well as about the characteristics of changing the values of the parameters of their own elements in hierarchical structures. All this makes it possible to increase the level of objectivity in assessing the values of subjective probabilities and the degree of belonging to vague sets. However, even under these conditions, the greatest uncertainty is still the image created in the mind (or in the computer model) of the DM$^{(i)}$, containing a forecast of the development of the situation $S^{(N_1,\ I)}$ and assumptions about the actions of a competitor $U^{(N_2,\ I)}$. And it is on the basis of this image that the DM plans its actions, and if it has a lower level, then the actions of its subordinates. That is, the DM performs a sequence of actions, the mappings of which are given in dependence (3.9). If we take into account that the capabilities of an experienced DM$^{(i)}$ in the field of management sometimes amaze with their high level of quality, then for this case, it would not be a significant exaggeration to consider the planning of actions of such a DM close to optimal.

Indeed, an experienced DM intuitively and logically, based on the properties of his analytical thinking, training, psychological stability, practical experience, is able to solve such complex problems, a formalized solution of which is not available today. It is known that the validity of the decisions taken by an experienced LDPR, the higher the above-mentioned qualities of his personality, i.e. the higher the characteristics of his personal factor (PF).

For the informational approach, the explanation of this phenomenon can be constructed as follows.

The higher the PF characteristics of a given DM, the longer and more plausible the line of logical constructions it uses when determining the minimum coverage interval of the true value of any of the parameters it considers at the DM. This circumstance allows us to assert that there is a factor of objectification of subjective probabilities of degrees of belonging. At the same time, such a DM often does not operate (does not know) certain performance indicators, with the help of which, during optimization, its such successful actions would turn out.

He carries out his activities intuitively and logically. Therefore, in order to fully formalize the values of performance indicators $\overline{y_i}^{(N_1)}$, the solution of the inverse problem of the theory of optimal processes can be used: counting the plan $\overline{U}_i^{(N_1)}(v)$ at each stage optimal (or quasi-optimal), find the efficiency indicator $\overline{y}_i^{(N_1)}(v)$, in the sense of the optimality of which control is obtained $\overline{U}_i^{(N_1)}(v)$.

In the same way, it is possible to formalize the performance indicators of the opposing side, if its actions are known—$\overline{U}_{ji}^{(N_2,I)}(v)$. According to these actions and known correlations and situations, it is possible to judge the motives, i.e. the indicators $\overline{y}_j^{(N_2,I)}(v)$. To implement such constructions, it is necessary to use AIS with developed data and knowledge banks capable of developing support for DMP in the current time mode for $DM_i^{(N_1)}(v)$. At the same time, such AIS should recreate a competitive chain—a model for a virtual $DM_j^{(N_2)}(v)$, his $\overline{U}_{ji}^{(N_2,\ I)}(v)$ and $\overline{y}_j^{(N_2,\ I)}(v)$, in order to objectify the support of the DMP of its DM.

4.4 Assessment of the Quality of DMP in Transport Processes and HAS

Any vehicle (HAS) has a purpose. Verbally, this goal can often be described in one word—prosperity. However, under different conditions, this word, as well as the goal itself, should be understood in different ways. In the conditions of wild capitalism, this is naked profit, and if DM_0 has broader views, his claims are much broader (see Fig. 3.2).

Previously, we defined these goals in the form of achieving the desired (for DM_0 and the entire company team) general condition of the vehicle in a certain future

$$S_{ij}(T) \in Y_{\text{is}}(n);\ n < T$$

while fulfilling resource restrictions and remaining in the legal field of activity:

$$u_i \in Y_{\mathrm{iu}}(n);\;\; S_{\mathrm{ij}}(T) \in Y_{\mathrm{is}}(n)$$

that is, the coordinating target designation of DM_0 should be determined from the condition:

$$U_{\mathrm{oi}}(n) \in \{Y_i(n),\ Y_{\mathrm{iu}}(n),\ Y_{\mathrm{is}}(n)\} \tag{4.15}$$

Then the current management should be evaluated by performance indicators characterizing the degree of deviation of the vehicle state (HAS) from its optimal value when moving this state from the current one at stage n to the final one—$S(T)$.

However, such an optimal "trajectory" of changes in the state of the system, as well as its defining parameters, has significant uncertainties. Therefore, as before (see 4.1, 4.2), it is necessary to introduce an entropy measure when assessing the quality of DMP in the vehicle (HAS). To do this, using a chain of consistent logical constructions, on the basis of objective information, a minimum coverage interval containing the true value of the desired parameter is selected. Then, on an objectively selected interval, the parameter value is intuitively and logically selected.

The application of this approach can begin with an analysis of the nature of uncertainty in determining any of the parameters included in the estimate of a priori entropy (4.2–4.5). Indeed, both in assessing the true values of the parameters of the state of the system S, and in evaluating the parameters of performance indicators, and, finally, in evaluating the parameters of the results of actions (i.e. the results of the DMP), it should be based on dependence (4.3). Extensive literature is devoted to evaluating the effectiveness of processes and analyzing its various aspects, most of which is given in the bibliographies of publications on this topic.

So, information plays a crucial role in DMP. However, ideas about the properties of information in time have not been used anywhere yet: information about the value of any parameter can be replenished by collecting additional data or by processing already collected data or become obsolete over time. Qualitatively, these processes can be represented by curves of changes in the values of the limiting entropy of estimation over time (see Fig. 4.3).

It is clear that the time spent on collecting additional data or processing them, for example, in order to obtain a forecast of data values, is directly proportional to the amount of information required for the DMP. Taking into account the costs of obtaining, processing and displaying information for the DM (or a computing complex that supports the DM) and the processes of its aging $H^{(\mathrm{cm})}$, it is possible to set a problem about the optimal value of processing time, i.e. about the maximum value of the amount of information or in terms of entropy:

$$H^*(P_s) = \min_t H(P_s) = \min_t \left[\sup\ H(P_s) + H^{(\mathrm{cm})}(P_s)\right] \tag{4.16}$$

The values of such periods of information delay depend both on the methods of obtaining–measuring data, and on the methods of its processing–forecasting.

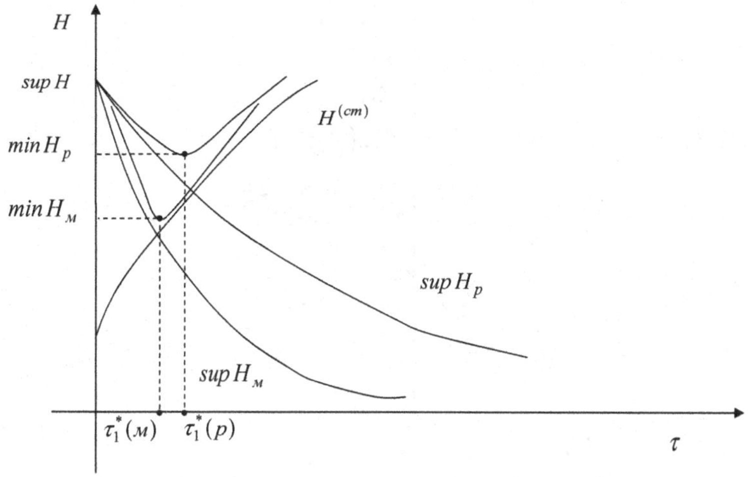

Fig. 4.3 Change in entropy when additional data is obtained or when summing predicted and aging, already available data

Taking two extreme methods: manual and machine, we get two extreme optimal values: $\tau^*(P_S)_p$ и $\tau^*(P_S)_M$ (see Fig. 4.3). At the same time, it is clear that $\tau^*(P_S)_p > \tau^*(P_S)_M$ by $\sup H(P_S)_p \geq \sup H(P_S)_s$, and then we will get different values of the residual information.

Usually, in practice, the DM determines the period $\tau^*(P_S)$ intuitively and logically, comparing it with the minimum amount of information that is achievable and necessary for DMP. At the same time, of course, there remains an unimproved value of uncertainty, the value of which is estimated as residual entropy:

$$H^*(P_s) = -\sum_{s=1}^{r} \log P_\rho; \left(\rho = \overline{1,\ r}\right) \tag{4.17}$$

where P_ρ—the probability of the true value ρ—parameter falling into the coverage interval is equal to. The corresponding coverage interval for this "strong-willed" case on the part of the DM is called the minimum—$\mathfrak{I}$(min).

To answer the question: is the information obtained in this case sufficient for the DMP, it is necessary to link the values of the amount of information with the purpose pursued by this vehicle (HAS), see (4.15), i.e. we are talking about the value of information.

At the same time, if the goal is achievable and its quantitative measurement is available (profit, final distance, hit accuracy, etc.), then the value of information in the conditions of confrontation, for example, two vehicles (HAS), can be measured in the form of an increase in the efficiency indicator with "hints" scouted about the enemy's strategy (see Fig. 4.4)

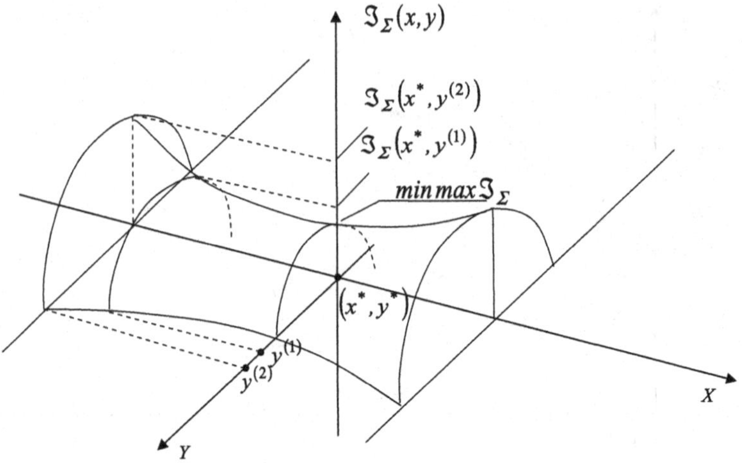

Fig. 4.4 Change in the performance indicator when changing the selection data when making decisions of the opponent in the matrix game.

$$\Delta\Im_\Sigma = \Delta \mathrm{I}\left(\overline{U}^{(x)}\right)$$

In case of unlikely achievement of the goal, it is convenient to measure the value of information in the way proposed by A. A. Harkevich. Then the measure of the value of additional information will be the logarithm of the quotient of the probability after it is received:

$$\text{Ц}(I) = log\frac{P(I_0)}{P(I_0 + \Delta I)}. \tag{4.18}$$

At the same time, it should be understood that the value can also be negative—in the case of misinformation.

So, the coverage interval corresponding to the entropy value $\min_t\left[\mathrm{Sup}\ H_0 + H^{(\mathrm{cm})}(t)\right]$, that is, the value $\tau^*(P_S)$ from (4.17), it can be considered the largest for each of the parameters of the state, the composite indicator of efficiency and the results (aftereffects) of the actions of the warring parties $\overline{L}^{(\max)}$. The best option, indicating the high qualification of the DM, is the case of $\overline{L}^{(\max)} \approx \overline{L}^{(\min)}$.

In general, the value of the unimproved entropy (4.16) serves as an objective measure of the quality of DMP. A relative assessment of the quality of the DMP can be considered a value that determines the degree of reduction of the initial uncertainty at the time τ^*:

$$\varphi(\text{Ц}, \Delta I, \Delta\Im_\Sigma) = 1 - \frac{H^*(P_\Sigma, \tau^*)}{sup\, H(P_\Sigma, \tau_0)}. \tag{4.19}$$

If you enter the amount of allowable losses in the form of, for example, the accuracy value determined by the indicator $\delta\mathfrak{I}_\Sigma$, i.e.

$$\left|\mathfrak{I}^* - \mathfrak{I}_{\Sigma\partial ocm}\right| \leq \delta\mathfrak{I}_\Sigma, \tag{4.20}$$

then the probability of choosing the optimal DM from the interval estimated intuitively and logically $L_{\mathfrak{I}_\Sigma}$ equal with an equally probable distribution

$$P\{\left|\mathfrak{I}_\Sigma - \mathfrak{I}_\Sigma^*\right| \leq \delta\mathfrak{I}_\Sigma\} = \frac{2\delta\mathfrak{I}_\Sigma}{L_{\mathfrak{I}_\Sigma}} \tag{4.21}$$

The limiting entropy corresponding to a given DMP can be found as $H^*(\tau^*) = -\log P\{o\}$, that is, taking into account (4.21) and (4.22)

$$P\{\circ\}=10^{-H^{*(\tau)}}=10^{-\sup H[1-\varphi(Ц,\Delta \mathrm{I},\Delta\mathfrak{I}_\Sigma)]}. \tag{4.22}$$

Thus, the values of time τ^*, entropy values $H^*(\tau)$, the permissible loss of the efficiency indicator and the value of information are closely interrelated. Values $H^*(\tau^*)$, $\overset{*}{\varphi}(Ц, \Delta I, \Delta\mathfrak{I})$,, are consistent in probability with the performance indicator by which management is optimized U, that is, the DMP is carried out.

At the same time, $H^*(\tau^*)$ determines what uncertainty remains for the volitional decision-making on the part of the DM, i.e. the degree of its risk, by appointment τ^*, as a result, the value of the indicator deviates from the optimal by $\delta\mathfrak{I}_\Sigma$.

Relying on the phenomenological properties of human thinking, one can try to explain the fact that in the DMP, there is not a "simple" search for solutions or any other (e.g. simplex or the method of the fastest descent) method of formalizing the search for the necessary information, but such a phenomenon as associative memory is included. We can talk about "recognition", "recognition" of such features that allow the DM to rank information by time and by stages of the DMP, i.e. to think in terms of the dynamic theory of information. This allows us to put forward one of the important rules in the HAS DMP—the rule of sequential elimination (reduction) of uncertainty. Thus, if you also introduce a fee for information into the performance indicator, then not only does the information age over time, it also becomes "more expensive" or, at least, does not fall in price on the scale of such an indicator.

Human thinking-DM has selectivity and generalizing property. It is characterized by two basic principles of complex systems: decomposition and aggregation, i.e. search from general to particular in the absence of a complete search, but under conditions of selective choice of information (associativity).

This allows an experienced DM to carry out a DMP close to optimal. However, it should be noted that this proximity significantly depends on the complexity of the tasks that form the problem of optimizing the DMP in the conditions of confrontation.

It is possible to introduce some measure of the complexity of the tasks that form the problem of DMP, as a function of the number of independent variables—factors that determine the conditions and models of HAS activity.

Such a function or the very number of independent variables simultaneously taken into account by the DM (or VC—when automating the support of the DMP) can be considered a function (number) of complexity. It is clear that in the HAS, the higher the control element in the hierarchy, the lower the value of such a function, because the more complex the functioning process (with a larger number of hierarchy levels) and the less time for decision-making, the fewer factors and conditions are taken into account by the DM, i.e. the complexity limit decreases with the growth of the hierarchy.

At the same time, it is clear that such a limit is a non-decreasing function of the time used for the DM; i.e. such a limit can be determined after a certain limit amount of information simultaneously used by the DM. And this amount of information is determined by the limiting a priori entropy (from 4.3)

$$\lim H^{(i)}(\Phi_S) \tag{4.23}$$

where (i) is the level of generalization—hierarchy; Φ_S—the number of complexity is the number of factors and conditions that are extremely taken into account.

With a decrease in the degree of generalization, that is, with less aggregation (when moving from one stage to another), the number of tasks that need to be solved with DMP increases.

The significance of the management selection criteria, i.e. alternatives to decision-making, also varies from stage to stage. All this increases the complexity. But the condition should follow from (4.23)

$$\sum_{l=1}^{L(i)} \lim\ H_l^{(i)}(\Phi_S) \geq \sum_{l=1}^{L(i)} H_l^{(i)}(P_S),\ \ (l = \overline{1, L}), \tag{4.24}$$

where l—the number of the sequence of tasks to be solved at the i-th stage at the DMP; $L(i)$—the number of such tasks. For management in the vehicle as a whole:

$$\sum_{i=1}^{T} \sum_{l=1}^{L_i} \lim H_l^{(i)}(\Phi_S) \geq \sum_{i=1}^{T} \sum_{l=1}^{L_i} H_l^{(i)}[P_S(t)] \tag{4.25}$$

Thus, the entire vehicle (HAS) is able to function normally only if the a priori entropy does not exceed the permissible value of the maximum complexity of the problems as a whole, which, in turn, is determined by the value of the maximum a priori entropy, the number of hierarchy levels, the number of factors and conditions taken into account. And, if we also take into account some provisions of the dynamic theory of information, then, in addition to the structure (the number of hierarchical levels and the number of elements on each of them), the functioning of the TC (HAS) is also determined by the value of the time of the DMP at each of the levels and for each of the elements.

If the value $\tau^*(P_S)$ the DM is defined and the condition is met $\overline{L}^{(\max)} \approx \overline{L}^{(\min)}$ for the parameter vector P_Σ, then it can be argued that in this case the definition of a rational decomposition, i.e. the structure of the vehicle is found from the condition of minimizing the time of DMP.

4.5 Informatization and Automation of DMP in the TSy Abroad

The use of various means and systems of informatization and automation of DMP at various stages of the functioning of the vehicle (organization of transport space, organization of traffic flows, direct control of the movement of vehicles) is necessary to increase the reliability of the human operator, to ensure the specified levels of technological safety of transport processes, as well as regularity and acceleration of traffic flows. Specific methods of implementation and options for building such systems for different modes of transport may have significant differences, but the goal of their implementation and use is always to ensure the safety and economic efficiency of the functioning of transport systems. In addition, at the present stage, the direction of development of universal (from the point of view of the mode of transport) and global (from the point of view of the possibility of use in various parts of the globe) systems and means providing such processes as search and rescue, navigation, communication, observation of the actual trajectories of vehicles, provision of meteorological information and etc.

It is clear that such systems are based on the use of satellite technologies, and first of all, they are used in aviation and marine modes of transport:

GNSS—Global Navigation Satellite System;

GPS—Global Positioning System;

AFSS—Aviation Fixed Satellite telecommunication Service;

GLONASS—Global Orbital Satellite System;

INMARSAT—International Maritime Organization for Mobile Satellite Communications;

COSPAS-SARSAT—international satellite system.

New standards and requirements for means and systems of informatization and automation are being introduced in other modes of transport. For example, in Europe, the IRIS quality standard for the railway industry has been put into effect, based on the ISO 9001 international quality standard. The new standard takes into account the specifics of railway transport and defines the quality requirements for traffic controls, signaling devices, translation mechanisms for switches and the procedure for certification of their manufacturers. In terms of its parameters, this standard is similar to the one already in force in the aviation industry.

In general, in the EU member states, much attention is paid to the introduction of the latest automation and informatization systems on railways. One of the most important measures in this area is the phased implementation of the European Train Traffic Management System (ETCS), which provides for several levels of automation.

For example, in May 2006, work was completed on equipping high-speed passenger trains with ETCS Level 1 drive aggregates on the Madrid–Zaragoza line. This will allow the daily use of 8 trains with a speed of 250 km/h. The planned implementation of the second level of this system will bring the speed of trains to 385 km/h.

For the first time, the European Level 2 Train Traffic Management System on a public railway in passenger traffic was introduced in December 2005 in Germany on the Peterborg–Leipzig section. By 2010, in connection with the creation of the Trans-European Transport Corridor in Germany, it is planned to equip 4500 km of public railway tracks with such a train control system.

The introduction of the European automated ETCS system of the second level is also envisaged within the framework of the program for the creation of a high-speed train network in Italy. The implementation of this program will allow connecting the largest centers of the country—Rome, Turin, Salerno, Naples, etc. ETR 500 trains with a speed of 300 km/h have been provided on the Rome–Naples section with a length of 216 km, which has already been put into operation.

Another important area of automation in railway transport is the electronic centralization of arrows and signals. The company Siemens Transportation Systems has completed work on the creation of a new electronic centralization system for arrows and signals (ESTW) at the main station Frankfurt am Main (Germany). This system is modular, covers 25 tracks at the main station and is integrated into the train traffic control post. An important element of the decision-making support process in the implemented system is the possibility of widespread use of simulation modeling by the station attendant.

The new ESTW electronic system will also be implemented as part of the modernization of the electronic centralization of arrows and signals on two lines of the Hamburg metro.

As for the automation of monitoring processes in order to control the movement of transport facilities, an example of a joint German-Swiss development of a new computer system for monitoring the movement of trains on public railways "e-train" can be cited. The operation of this system is based on the use of a global vehicle location system using satellite communications (GPS) and GSM mobile communications.

The introduction of energy-saving technologies can be singled out as another direction of automation. An example of the introduction of energy-saving technologies in the movement of trains in Germany is the automated system "Locomotive driver's Assistant".

"ENAflex-S". It should be noted that the work of this system is based on the extensive use of modeling statistical information and possible conflict situations.

Thus, today a significant number of automated systems and automation devices (ETCS, ESTW, automatic locking systems, etc.) are used on the railways of Europe.

However, their effective use implies, among other things, ensuring their interaction, which requires the creation of special interfaces and data exchange protocols.

Again, for German railways, the SAHARA system has recently been created, which is a universal interface and data exchange protocol that can ensure the interaction of various automated systems and information technologies with a high degree of reliability.

Also in Germany, an electronic system for diagnosing and ensuring traffic safety in railway transport (ESDIS) has been created, which monitors the technical serviceability of structures and automatically sends an alarm signal in case of a malfunction. The system can be integrated into the post of electronic centralization of arrows or into the post of dispatching control of train traffic.

As part of the introduction of new information technologies, including in order to improve the level of service, German railways, as an experiment, equipped 12 trains on the ICE 3 class high-speed train line with a wireless Internet connection system via a local WLAN network or through its own proprietary Intranet local system. The connection is made via UMTS radio communication. In case of positive results of the experiment, a decision may be made to extend such a system to further sections of high-speed train traffic.

And in March 2006, in Germany, at the 25th Scientific and Practical conference on informatization in transport, the issue of the introduction of electronic fare payment technologies on railway, urban and other modes of transport (eTicketing) was discussed.

In another part of the world—China—in accordance with the program of modernization and development of transport in connection with the preparation for the 2008 Summer Olympics, modernization of the Beijing subway is planned. As a result, it is planned to implement the URBALISTM train traffic management system based on the use of a local wireless WLAN network with radio communication and an electronic system for the centralization of arrows and signals. The system includes a subsystem for automatic control of the ATO train movement, which is capable of optimizing such control.

In addition, interesting solutions have recently been introduced to automate and informatize the processes of organizing transportation and logistics. Appropriate systems and software tools allow you to support the decision-making process for such tasks as:

- ensuring cost-effectiveness, reliability and accuracy of order fulfillment when transporting goods between the logistics center and production buildings by means of driverless road trains and equipped with an automated RITA control system and transponders for route recognition;
- automation of warehouse management based on the universal software "LOGSTARJ", which is a further development of the software package "LOGSTAR";
- design of automated production and transport systems with three-dimensional visualization in the design process based on the software package "Smap-3D-Laoyt-Planning";

- automation of cargo traffic management and transport and warehouse operations based on the automated system "LFS 400";
- introduction of information technologies related to the identification of various goods and products based on the use of bar coding technology and radio frequency identification technology (RFID);
- automation of cargo flow management in internal logistics (WAMASMFR, etc.);
- optimization of the choice of containers when picking and packing orders in warehouses and logistics centers based on the ORionPI software;
- optimization of the use of warehouse space based on the automated system for determining the size and weight of goods "MultiscanPBM", which provides volumetric scanning and automatic weighing of goods;
- inventory management of goods in warehouses based on the "add*ONE" software;
- remote monitoring of the position and condition of specialized refrigerated and standard containers, vehicles based on mobile terminals of the SYMBOLMC3000 model.

In the Eurocontrol countries, uniform recommendations for automated air traffic control systems AS ATC are being developed. In accordance with these recommendations, ATC ACS should have such functions as Safety Nets (SNET)—security controls; Monitoring Aids (MONA)—controls; Medium-term Conflict Detection (MTCD)—detection of medium-term conflicts; System Supported Co-ordination (SYSCO)—automated coordination; Conflict Resolution Advisor (CORA)—recommendations for conflict resolution; Arrival Manager (AMAN) is a system for organizing the incoming flow of aircraft; Departure Manager (DMAN) is a system for organizing the departing flow of aircraft.

The SNET functions (security controls) provide a warning to the dispatcher about potentially conflicting situations of aircraft (short-term conflicts), about the danger of lowering aircraft below the minimum safe altitude, as well as about the possible penetration of aircraft into restricted areas of airspace use.

The MONA functions (controls) ensure the detection of deviations of aircraft from the planned trajectories of movement and provide assistance to the dispatcher in controlling aircraft by providing him with information about real or predicted deviations from the planned trajectory (e.g. when the aircraft deviates beyond the boundaries of the air corridor). In addition, MONA reminds the dispatcher of the need to perform certain technological operations, actions.

The Medium-term Conflict Detection (MTCD) functions are a planning tool and provide conflict detection in the range from 0 to 20 min. MTCD provides dispatchers with information to help them resolve conflicts. One of the main means of MTCD is the conflict and risk display (conflict and risk display), which displays the time remaining before the conflict and the distance of the minimum divergence of aircraft.

SYSCO (automated coordination) functions provide automated coordination between adjacent air traffic control sectors. Coordination is provided via an on-screen interface.

The conflict resolution recommendations (CORA) functions provide support for decision-making on conflict resolution. Currently, the requirements for CORA are at the development stage.

AMAN (the system for organizing the incoming flow of aircraft) provides the dispatcher with recommendations on the organization of the order of incoming aircraft in order to ensure maximum runway capacity.

DMAN (the system for organizing the departing flow of aircraft) provides the dispatcher with recommendations on the order and time of starting the engines of departing aircraft in order to ensure maximum runway capacity.

The Russian company CJSC VNIIRA-ATS is currently developing and supplying a unified set of ATC ATC "SYNTHESIS", which complies with the standards and recommendations of ICAO and Eurocontrol.

Based on the presented brief overview of foreign means and systems of informatization and automation of vehicle processes used in decision-making at different levels and stages of vehicle operation, it can be concluded that the future belongs to systems offering a comprehensive solution to the problems of ensuring safety requirements, economic efficiency, improving the level of service, as well as logistical tasks.

Chapter 5
Modeling of Transport Processes in the Optimization and Functioning of the Transport Space

5.1 General Characteristics of the Tasks of the Organization and Functioning of Transport Space Systems

The most characteristic tasks of the organization of transport space for almost any type of transport are the tasks of choosing a network of traffic routes, rational placement of logistics centers, as well as points of placement of service stations, assessment of the required characteristics of the capacity of all elements of the transport space in conditions of ensuring security restrictions. Mathematical models of transport processes are most often models of network graph theory, queuing, mathematical programming in the form of linear programming transport problems, integer programming and other optimization problems.

It should be noted that most of the transport processes in the organization and functioning of the transport space in various modes of transport do not have a single well-founded and formalized assessment—an indicator of efficiency, as well as mathematical models that are uniform in terms of the coverage of factors. All this often forces at least some general principles to be taken into account when modeling such processes. At the same time, the type of implementation of such accounting is chosen separately each time: in the form of constraints (mathematical inequalities—most often), variations of parameters (areas of acceptable solutions) and convolution of indicators (weighted sum, quadratic or minimax). Similar general principles include:

- the principle of "equal strength", i.e. the construction of models that allow you to obtain organizational solutions that lead to a uniform possible load on the elements of the transport space, minimizing congestion, traffic jams, peak loads, etc.;
- the principle of an integrated approach to modeling, when all possible factors, conditions and mutual influences and interrelations of transport processes are taken into account in the organization of transport space;
- the principle of a guaranteed approach, when modeling in the presence of uncertainty, they count on the possible best option of the situation.

© The Author(s), under exclusive license to Springer Nature Singapore Pte Ltd. 2023

G. A. Kryzhanovsky et al., *Modeling of Transportation Aviation Processes*, Springer Aerospace Technology, https://doi.org/10.1007/978-981-19-7607-0_5

Many transport processes in determining the main characteristics of the transport space, such as, for example, the number of traffic lanes, the number of waiting areas in the airspace of an air hub, the number of runways (runways) of an airfield, the number of berths of seaports, can be represented as the processes of servicing incoming and outgoing vehicles. These and many other examples of transport processes are then modeled in terms of queuing theory (QT). If we consider such processes to be Markov processes with varying degrees of approximation, then it is possible to use already sufficiently fully developed QT methods to determine the main parameters of the elements of the transport space. The main requirement in this case is the requirements for the proximity of the characteristics of the flows of vehicles falling on the elements of the transport space to the Poisson random process. The most interesting types of queuing systems (QS) modeling transport processes are QS with waiting (with queue) limited and unlimited, with failures, single-channel and multi-channel, etc. The most important set or desired characteristics of processes in such QS are then the number of channels—n (lanes, waiting areas, runways, berths, etc.), the intensity of the flow of applications (arriving or departing vehicles, service requests, etc.)—$\lambda(t)$, the average number of requests served per unit of time by each of the i-th channels—$\mu_i(t)$, conditions for the formation of a queue. Then for the simplest case $n = 1$, $\lambda = \lambda(t)$, with application service during T_r—random time distributed according to the exponential law:

$$f(t) = \mu \cdot e^{-\lambda \cdot t}, \quad (t > 0) \tag{5.1}$$

relative throughput (by $\lambda = \text{const}$ and $t \to \infty$):

$$q = \frac{\mu}{\lambda + \mu} = \frac{1}{\rho + 1}; \quad \rho = \frac{\lambda}{\mu} \tag{5.2}$$

The absolute throughput is then defined as:

$$A = \lambda \cdot q = \frac{\lambda \cdot \mu}{\lambda + \mu} \tag{5.3}$$

Similar dependencies for QS with waiting, and a different number of channels, and the number of places in the queue, as well as dependencies (5.1–5.3) are well known and can be found in the literature.

As an example, the organization of transport space in air transport can be cited. The basis of the organization's processes are the processes of organizing air traffic control and making organizational and structural decisions.

Taking into account the basic principles in this case, although it is not equivalent to specific methods, which means that it does not allow finding the best solutions to problems for the processes of the ATC organization stage, it makes it possible to obtain many solutions belonging to the range of acceptable values, which is a significant example.

In accordance with the principle of "equal strength", i.e. ensuring a uniform load on the elements of the ATC system, models are built and organizational and structural decisions are made when organizing ATC. The structure of the ATC system includes control rooms that directly control the movement of aircraft in all areas of the space controlled by it. This is how the simplest ATC contours are formed (see Fig. 3.3), consisting of a number of elements, the most important of which is the dispatcher himself. The dispatcher's capacity $\mu_i(t)$ and permissible workload $m_i(t)$ are determined by the psychophysiological characteristics of the human operator and the capabilities of the technical means and ATC systems that he uses.

In real conditions, the workload of ATC zones with the number of aircraft $N_i(t)$ may vary significantly and exceed the permissible workload of the dispatcher $m_i(t)$. In cases when the ATC area is overloaded, $N_i(t) > m_i(t)$. This situation occurs when the intensity of air traffic $\lambda_i(t)$ exceeds the bandwidth $\mu_i(t)$ this zone (dispatcher). When overloading, the probability of errors and failures in the dispatcher's work increases, the reliability of the simplest circuit of the ATC system decreases, which reduces the safety of air traffic.

Therefore, at the stage of ATC organization, it is necessary to apply mathematical models that allow to obtain organizational solutions that exclude overload of individual elements of the simplest circuit and ensure uniform load distribution between them, in which $\mu_1(t) \approx \mu_2(t) \approx \ldots \approx \mu_n(t)$ and the conditions are satisfied:

$$\lambda_1(t) \leq \mu_i(t);\ N_i(t) \leq m_1(t)$$

that is, the optimization problem of the form is solved

$$I = F\left[\sum_{v=2}^{\mu+1} \omega_v (\mu_v - \mu_{v-1})^2\right] \to \min \tag{5.4}$$

by $N_i(t) \leq m_i(t);\ \lambda_i(t) \leq \mu_i(t)$

The physical essence of this task consists in the maximum possible smoothing of the throughput of elements of the simplest contour and adjacent contours.

Organizational decisions taken in accordance with this principle include: dividing the airspace controlled by the ATC system into zones and determining the volumes (sizes) of these zones; creating a network of control rooms in accordance with the statistical characteristics of aircraft flows; automating and improving the processes of direct ATC in zones in order to increase the permissible workload dispatcher $m_i(t)$ and the bandwidth of the zone $\mu_i(t)$, separation of functions performed by one dispatcher between two in the structure of the same point or between the dispatcher and the operator.

For example, in the RP in the RDS with a high traffic intensity of aircraft, the functions of the RDP dispatcher (RC) are performed by dispatchers: radar control and procedural (graphical) control. This allows you to evenly distribute the traffic control load in this area and ensure reliable operation of the ATC system.

The principle of a guaranteed approach (solutions) is used in the analysis of operations, tasks and processes of the ATC organization stage in conditions of uncertainty. The essence of this principle is to find guaranteed solutions to problems and research operations in processes at the stage of the organization of ATC in such conditions with the expectation of the worst case. At the same time, in each task, the uncertainty condition is modeled by assuming the existence of another participant in the study with opposite interests. Such an approach to the study of operations and solving problems related to the processes of the ATC organization stage is called a game approach or an approach that allows you to get guaranteed solutions for the worst conditions. The methodology of solving game problems is based on game theory, which explores the issues of decision-making in conditions of uncertainty.

By their nature and nature of movement, controlled aircraft as dynamic objects cannot be stopped during flight. The flows of control, working (via feedback channels) and coordinating information about air traffic within the boundaries of ATC zones are discrete in nature and should ensure the interconnection of elements of the simplest contour. In addition, the flows of aircraft (especially incoming ones) in areas with high traffic intensity are characterized by the Poisson distribution law, i.e. they require an almost continuous process of operational ATC in time. All these factors lead to the need to organize a system of continuous monitoring of the movement of aircraft within the boundaries of the zones and the continuous operation of traffic control points of the traffic service and the entire ATC system as a whole. It is for this reason that the results of QT and modeling of transport processes using the conclusions of the QS are widely used.

5.2 Modeling of the Processes of Placing Elements of the Transport Space

When organizing a transport space, in most cases, rational decision-making on the choice of the location of an element of the transport space (ETS) is essential for any type of transport. When modeling the processes of organizing transport space, the most important tasks are the construction of such models of rational choice of ETS locations, for example, sea ports, air transport, automated logistics centers, navigation equipment and many other examples of ETS placement.

One of the options for mathematical modeling of problems of this kind is the use of linear programming. Using again the processes of organizing air traffic control as an example, it is possible to obtain models of the organization of the control of the movement of aircraft in the form of models of optimal placement of surveillance radar stations (SRS).

The meaning of the task of rational placement of the SRS is that, having initial data on the boundaries of the considered ATC zone, the characteristics of air traffic in it, tactical, technical and economic data on the ORLS, to find such points of installation of the ORLS, at which radar control of the movement of aircraft will be provided at the lowest cost for the purchase, installation and operation of the SRS. Such a task has several modifications that take into account the requirements of continuous or selective coverage of various areas of the ATC zone by radar control, the presence or absence of previously installed radar stations and much more. When constructing mathematical models and their modifications, the formulation of the rational placement problem can be carried out in many ways. At the same time, its formulation in the form of an integer linear programming problem is the most constructive. Then its practical implementation can be carried out in the following way.

Suppose some ATC zone is specified in which it is necessary to create a control system using radar. With the help of a uniform grid, it is divided into fairly small sections, the size of which (grid pitch) is selected based on considerations of taking into account the errors in the construction of the SRS zones, the accuracy of economic and other data, and is usually equal to 20 km. The grid nodes form a set $M = \{1, 2,\ldots, i,\ldots,m\}$ points, and in each of the subdomains Ω_j ATC zones where flights with different characteristics are carried out, there are needs for their own type of control, and therefore.

$$M = \bigcup_{j=1}^{n} M_j,$$

where M_j—subdomain points Ω_j.

In the subdomain Ω_j, the set is also given $N = \{1, 2,\ldots,n\}$ points of possible radar installation locations. The sets M and N generally do not coincide, since usually the number of points N is significantly less than the number of points of the set M.

Thus, for each j-th point belonging to the set N, it is characteristic to provide control with a given property. Different types of control require the placement of different types of SRS.

So, if in the subdomain Ω_j flights of heavy types of aircraft are carried out, and in Ω_k, for example, only PANH flights, then, of course, both the control requirements and the types of radar in Ω_j и Ω_k must be different. Let the weight coefficients be assigned to each of the available types of SRS $\omega_j^{(s)}\left(s = \overline{1, p}\right)$, reflecting, for example, the total costs of purchasing, installing and operating an s-type radar on site (at a point) located in the subdistrict Ω_j. Let us introduce the matrix $\{a_{ij}\}$:

$$a_{ij} = \begin{cases} 1, & \text{if the point } i \text{ is included in } M_j; \\ 0 - & \text{in the opposite case.} \end{cases}$$

The Boolean variables will then be searched for

$$x_j^{(s)} = \begin{cases} 1, & \text{if the set } M_j \text{ included in the desired coverage;} \\ 0, & \text{in the opposite case.} \end{cases}$$

Then $\sum_{s=1}^{p}\sum_{j=1}^{n} x_j^{(s)}$—determines the number of sets of all types of control included in the coverage, and the weighted number of SRS is then determined by the indicator

$$I = \sum_{s=1}^{p}\sum_{j=1}^{n} \omega_j^{(s)} x_j^{(k)} \tag{5.5}$$

The presence of a control of the types is determined by the sum $\sum_{s=1}^{p}\sum_{j=1}^{n} a_{ij}^{(s)} x_j^{(s)}$, characterizing the number of sets included in the coverage and containing the i-th point. Then the condition for the presence of coverage has the form:

$$\left.\begin{array}{l} \sum_{s=1}^{p}\sum_{j=1}^{n} a_{ij}^{(s)} x_j^{(s)} \geq 1; \\ x_j^{(s)} = 0,\ 1; \\ \left(s = \overline{1,p},\ i = \overline{1,m},\ j = \overline{1,n}\right) \sum_{s=1}^{p} x_j^{(s)} \geq 1 \end{array}\right\} \tag{5.6}$$

The task boils down to choosing such $x_j^{(s)}$, under which the exponent I takes the smallest value and the conditions are satisfied (5.6).

If we consider the control needs to be homogeneous in the first approximation, then $p = 1$. Taking into account approximately the same characteristics of the viewing areas for the route radars, it can be assumed that the power (number of points) of subsets of M corresponding to the viewing area Ω_j is determined only by the nature of the terrain and air traffic in the vicinity of the selected conditional location of the SRS installation. In this case, the problem of rational placement (5.5), (5.6) will be simplified and will have the form:

$$I = \sum_{j=1}^{n} x_j \to \min \tag{5.7}$$

$$\left.\begin{array}{l} \sum a_{ij} x_j \geq 1; \\ \left(i = \overline{1,m},\ j = \overline{1,n}\right) x_j = 0,\ 1 \end{array}\right\} \tag{5.8}$$

The formulation of the problem of rational placement in this form or formula for a large-sized ATC zone, including several departments or any regional part, the European part of the Russian Federation or the entire territory of the country, leads to large dimensions of the matrix $\{a_{ij}\}$. So, for the area Ω, including Penza, Kazan, Ufa, Orenburg and Kuibyshev districts of the dispatch service, the dimension of the matrix $\{a_{ij}\}$ 491 × 24.

The significant dimensionality of the conditions of the initial matrix in the rational placement problem forces us to turn to proven techniques based on the principle of decomposition, the essence of which consists in: splitting a complex problem into a number of simpler ones, the solution of which is foreseeable and can be obtained relatively simply; developing a system of interrelated solutions to simple problems (by introducing parameters into constraints, weight coefficients in performance indicators, additional unknowns and other techniques); obtaining estimates indicating the proximity of the synthesized solution combining solutions of simple problems and the original complex problem.

The application of the decomposition principle is most simple and effective in cases when splitting arrays of expressions of the initial conditions of the problem does not lead to distortions of its physical nature. Such a case arises when splitting the matrix of the initial problem of rational radar placement. Decomposition of this problem for the domain Ω a large area is carried out in the following sequence: build a matrix $\{a_{ij}\}$ of the initial radar coverage for the entire area under consideration; to divide the resulting matrix into submatrices, the dimension of which allows performing the operations of the solution algorithm; to perform these operations in each of the submatrices; to combine the results into a new general matrix of conditions; and to perform the operations of the solution algorithm with a new general matrix.

It follows from the above that a significant place in solving the problem is occupied by the preparation and quality of the source data. Their totality includes types of SRS, restrictions (prohibitions) on the placement of SRS, of a physical, geographical, demographic and departmental nature, places of possible installation of SRS on the territory of a given zone and near its borders, a network of existing and planned routes, as well as their minimum echelons for building subdistricts requiring continuous radar monitoring.

For the selected locations of the proposed radar installation, viewing zones are built taking into account the minimum flight echelons, closing angles created by the terrain. After that, a set of points is formed, displaying the subdomains of continuous monitoring. Using this data, it is possible to build a radar coverage matrix $\{a_{ij}\}$, which is the starting point for solving the general problem.

The algorithm for solving the problem of rational placement of the SRS can be considered as universal, since it allows, with minor modifications, to solve the problems of rational placement of various types of radio engineering and other means of navigation and ATC, such as RSBN, DRL, radio direction finders, HF communication stations and laser trajectory meters. Of the many exact and approximate methods for solving integer linear programming problems (5.5), (5.6) and (5.7), (5.8) with a significant dimension (n >> 10), branch and boundary methods, clipping methods and heuristic methods are applicable in practice. The proposed algorithm can also be attributed to a composite algorithm that includes a heuristic and a branch and boundary type algorithm for constructing a set of minimal coverings and analyzing the obtained solutions—minimal coverings in order to select the most rational of them.

To calculate the lower estimate (in the minimization problem) of the efficiency indicator reflecting the weighted number of SRS included in the minimum coverage, the first branch of the solution is constructed, where columns with the highest norms are selected in the matrix of conditions for each generation. This makes it possible, taking into account the peculiarities of the matrix structure of the initial and minimized coatings, to assert that the probability of obtaining a minimum coverage when implementing such a branch is almost equal to one. Otherwise, if it turns out that the obtained estimate does not give a minimal solution, then the algorithm allows you to refine the estimate and ensures the detection of all minimal coverages. The composite algorithm includes two algorithms, one of which provides a solution to the problem of determining the set of minimal coverings, and the other—analysis and selection of a rational solution.

The first algorithm involves performing a number of operations. At the same time:

- An array of source data is formed. The matrix plays a special role here $\{a_{ij}\}$, the rows of which are formed by the numbers of the points of the coverage area, and the columns by the numbers of the points of possible radar installation locations.
- The SRS that has already been established and fixed by the condition of the task is excluded. To do this, their numbers are in the matrix $\{a_{ij}\}$ excluded.
- The number of radars in the initial coverage is set I_k and a comparison is made with the one obtained at this iteration, which allows you to cut off the branch of the solution search.
- The norm of each SRS is calculated, i.e. the number of points covered by the control.
- The SRS with the highest rate is selected. This allows this branch of the solution, using the radar with the highest norm with a probability close to one, to obtain a minimal solution, which significantly reduces the amount of calculations due to the separation of other branches.
- The condition is checked:

$$\left[\frac{N_k}{\max[\text{NORMA}(N_k)]}\right] + (N_k - 1) > I_K,$$

where square brackets denote the operation of determining the nearest larger integer, and the left side of the inequality characterizes the minimum number of iterations that can lead to a solution, i.e. the minimum number of SRS that can be in this solution. Then move on to the next iteration $N_k = N_k + 1$;

- A new array of EAGLES lists is formed by excluding from them the SRS number included in the solution of this branch in the previous iteration.
- The number of lists is calculated, i.e. the number of SRS at this iteration N_k.
- The condition of the end of the solution for this branch is checked $N_k = 0$.
- The number of radars in the resulting solution is compared with I_K. Almost always I_K it turns out to be big. If not, then the previously obtained solutions are excluded and I_K assign a new higher value.

- A new solution is recorded in the list of required minimum coatings. To the previous iteration $N_k = N_k - 1$, they switch to check whether it is possible to get a new one from the previous solution by replacing only the last number. If this cannot be done, a return is made for another step.
- The condition for considering all branches is checked by comparing the iteration number with zero. A positive answer here means that all branches are considered and the set of minimal covers is found. If this does not happen, then the cycle repeats from a new calculation of norms and further.

The choice of the optimal solution from the resulting set of minimal solutions is determined in practice by analyzing a variety of heterogeneous and contradictory requirements. Taking them into account by introducing a complex criterion is possible, but it is associated with significant difficulties caused by the non-stationarity of the requirements. In addition, when forming the matrices of the initial radar coverage, radar viewing areas are usually constructed using cartographic data, which does not always and does not fully allow taking into account the influence of local features. Therefore, a simpler and more acceptable method is to narrow down the list of possible coverage options by rejecting solutions, among which there are obviously no optimal ones. The final choice of the option is carried out with the involvement of additional data and expert assessments.

Rejection is usually carried out according to the degree of overlap of radar fields and the assessment of total economic costs. After the rejection of obviously unsuitable solutions, the rest (if there were more than one) are compared taking into account data on the proposed sites of installation of the radar, demographic and other features of specific areas. The experience of the solution shows that one of the variants of expert assessments is the most preferable here.

In order to take into account successful solutions to the problem of radar placement (and thereby gain experience in building a comprehensive performance indicator) obtained with the help of expert assessments, it is necessary to formalize the process of studying them. One of the variants of such formalization is the use of the inverse problem of the theory of optimal processes, as in the case of one of the variants of the construction of the performance indicator, when, according to a known decision taken for optimal, the efficiency indicator is restored. This allows experts to accumulate data on accounting for the most important factors. If we assume that rational placement is defined in the form of components $x_j^*\left(j = \overline{1, n}\right)$ vector X^*, then the task is set as follows:

Find the range of acceptable values for the weight coefficients ω_j performance indicator

$$I = \sum_{j=1}^{n} \omega_j x_j^* \tag{5.9}$$

provided that

$$\left.\begin{array}{l}\sum_{j=1}^{n} a_{ij}x_j^* \geq 1;\\ \left(i=\overline{1,m};\ j=\overline{1,n}\right);\ x_j^*=0,\ 1\end{array}\right\} \quad (5.10)$$

The solution of such an inverse optimization problem is found using Farkash's theorem, according to which it is possible to write:

$$\sum_{j=1}^{n}\omega_j x_j=\sum_{i=1}^{m_k}\sum_{j=1}^{n}a_{ij}^{(k)}x_j\lambda_i^{(k)}, \quad (5.11)$$

where $j=\overline{1,m_k}$– the number of equality constraints in (5.10); k– index of belonging to the k-th solution;$\lambda_i^{(k)}$—non-negative parameter.

Equality (5.11) is valid for every solution, so

$$\omega_j=\sum_{i=1}^{m_1}\lambda_i^{(1)}a_{ij}^{(1)}=\ldots=\sum_{i=1}^{m_k}\lambda_i^{(k)}a_{ij}^{(k)} \quad (5.12)$$

It follows from this expression that

$$\sum_{i=1}^{m_{s-1}}\lambda_i^{(s-1)}a_{ij}^{(s-1)}-\sum_{i=1}^{m_s}\lambda_i^{(s)}a_{ij}^{(s)}=0 \quad (5.13)$$

for ($s=2, 3, \ldots, k$) or in matrix–vector notation:

$$A\lambda=0,$$

where $A=\{a_{ij}\},\ \lambda=\left(\lambda_i^{(2)},\ldots,\lambda_i^{(k)}\right)'$.

Solving the system of algebraic Eqs. (5.13), it is possible to determine the components $\lambda_i^{(s)}$, which, in turn, will allow us to find the components from the expressions (5.12) ω_j. By applying this procedure several times, you can take into account the experience of the decisions made and narrow down the areas of setting coefficients ω_j in indicators of the type (5.9), (5.5).

The above algorithm is used to build projects of a radar control system for the movement of aircraft in the ATC zone of the district center in Arkhangelsk, a number of districts of the Volga administration and in other zones. As an example, let us consider some fragments of solving the problem of rational placement of the SRS for the Arkhangelsk regional center. The RC area is represented by a set of 347 points placed regularly in increments of 20 km horizontally and vertically. 36 localities have been selected as possible conditional locations for the installation of the SRS. By mapping a network of trails and other important elements of the organization for ATC, it would be possible to reduce (or increase) the number of points. Next, the

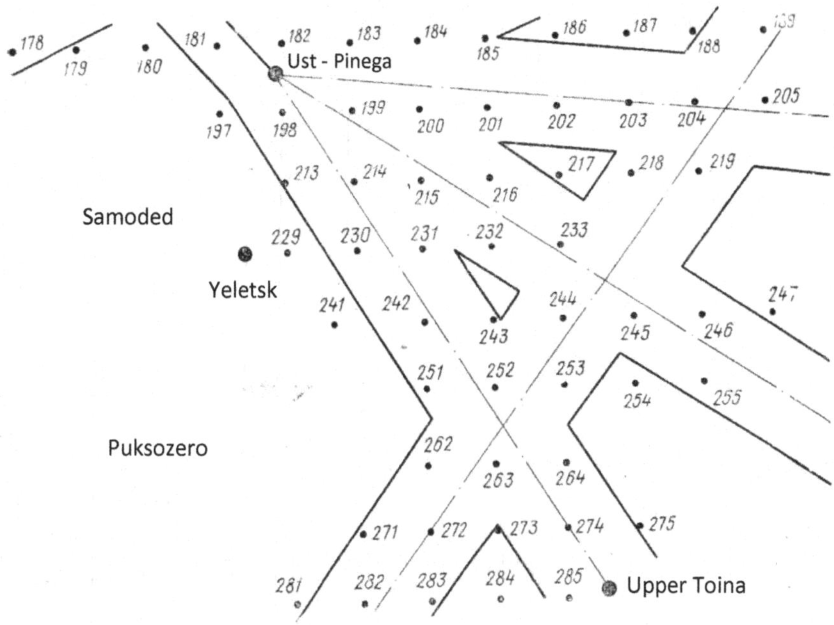

Fig. 5.1 Fragment of the network of routes prepared for the construction of the initial matrix

construction of the radar survey zones is carried out according to known methods or by approximate formulas. A fragment of the network of routes is shown in Fig. 5.1. With the help of the data obtained and taking into account possible locations of the SRS, the formation of the initial radar coverage matrix is carried out. Usually, the matrix is entered into the computer in the form of lists of SRS covering the g-th point of the set M_j, displaying subdomains requiring continuous radar monitoring in the area under consideration Ω.

An example of a fragment of one initial matrix minimized using an algorithm for minimizing its size is shown in Fig. 5.2. As an algorithm for minimizing the initial matrix, a method constructed using the well—known Yablonsky-Mcclassky dominance method can be used. Practically, this means the exclusion (absorption) of a number of identical rows, which corresponds to the case of covering a point of the region Ω the same combinations of SRS. Having excluded all but one of the identical rows in groups, they assign it a "weight" equal to the number of rows in the group, thus obtaining a smaller matrix. By taking into account the fixed SRS and repeating the row absorption operation, taking into account the change in the weights of the elements of the set M_j, a minimized matrix of the initial coating is obtained. Next, the algorithm described above is used.

The resulting solution allows us to propose a rational design of a system for controlling the movement of aircraft in this ATC zone. As follows from the above, a significant part of the operations of the process of creating such a project can be automated using computer. The whole set of tasks and operations that form the

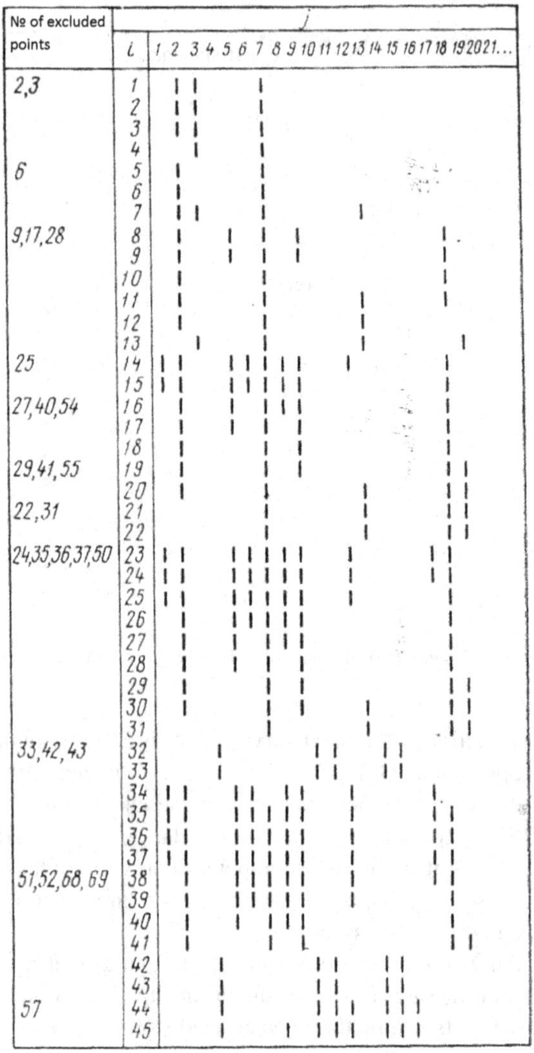

Fig. 5.2 Initial radar coverage matrix

process of creating a control system is carried out using a complex that includes both a computer and a group of experts who are well aware of local conditions and the specifics of this ATC zone, for which the issue of creating a control system is being solved. Therefore, we can talk about an interactive solution to the work of researchers with computer, i.e. only about the partial automation of such a process as the process of creating a control system. This circumstance is to a large extent characteristic of other processes of the ATC organization stage.

5.3 Infrastructure of Transport Systems. The Problem of Terminal Placement

The state of the transport systems of individual regions, entire countries and continents is determined not only by the state of the fleet of vehicles, but also, mainly, by the state of the transport space—transport routes, ports and terminals for each type of transport without exception.

The development of the economy of any of the countries is largely determined by this state, since the amount of contributions to the construction of main transport routes and to the construction of large ports (railway stations, airfields with terminals, river and sea berths with port terminals, automobile stations and cargo terminals, container terminal complexes, etc.) are comparable with the annual budgets of individual countries.

Examples are the implementation of the Baikal-Amur Railway construction project, the construction of a seaport complex in Nakhodka, the laying of a federal highway to the Far East, the construction of airport complexes in the Moscow Air Hub and the modernization of the entire air traffic control system using satellite technologies, and many other ongoing projects in Russia.

The economic recovery is inextricably linked with the necessary quite tangible leap in the development of the entire infrastructure of the transport system. The development of the infrastructure of transport systems is associated with significant, one might say, global projects for the development of transport routes, transport ports and terminals, as well as communication, surveillance and management facilities.

It is advisable to show that the solution, for example, of such an important problem as the development of transport routes, begins with solving a problem called laying, and the problem of developing a network of ports and terminals and other transport infrastructure facilities—with solving the problem of rational placement, discussed in Sect. 5.2.

The simplest, but taking into account the requirements of practice, a variant of the problem gaskets can be represented as follows.

What is sought in this task is the type of function that defines the "profile" or "plan" (top view) of the transport path in two planes. Taking into account the dynamics of vehicles imposes a number of conditions on the appearance of this function.

So, you need to find a smooth function

$$U = f(x), \tag{5.14}$$

for which:

$$\left|\frac{df(x)}{dx}\right| \leq C_1, \tag{5.15}$$

where x—the axis of the laying route, U—deviation from the axis, C_1—the maximum permissible value of evasion, depending on the type of transport and vehicles used on this path

$$\left| \frac{\frac{d^2 f(x)}{dx^2}}{\left[1 + \left(\frac{df(x)}{dx}\right)^2\right]^{3/2}} \right| \begin{matrix} \leq C_2, \text{ if } \frac{d^2 f(x)}{dx^2} < 0 \\ \leq C_3, \text{ if } \frac{d^2 f(x)}{dx^2} > 0 \end{matrix} \tag{5.16}$$

where C_2 and C_3—the greatest values of the path curvature for the case of convex and concave curves $f(x)$,

$$U_i \begin{cases} \geq \overline{U}_i \\ \leq \overline{U}_i \end{cases}, \quad \overline{U}_i - \text{fixed points of the curve } f(x) \tag{5.17}$$

The specified (also selected as a result of solving the problem of rational route routing) profile of the transport path is determined by the function (and/or the value of a finite number of points connecting standard sections $f_j^*(x_i)$) kind of:

$$U^* = f(x^*) = \{f_1^*(x_k),\ k \in N;\ f_2^*(x_l),\ l \in N;\ \ldots;\ f_r^*(x_p),\ p \in N\}, \tag{5.18}$$

where $f_j^*(x_i),\ i \in N$: by $j = 1$—direct, when $j = 2$—circle, when $j = 3$—parabola, etc., starting at the point x_i and ending at the point x_{i+1}.

Then the task of laying reduces to the task:

$$\min_U I[U] = \int_{x_0}^{x_T} [f(x) - f^*(x)]^2 dx \tag{5.19}$$

under restrictions (5.14–5.18).

The solution of such problems can be found by dynamic programming methods or using the maximum principle of L.S. Pontryagin.

At the same time, it should be noted that the laying task, like the routing task, can also be formulated in the form of mathematical and sometimes linear programming problems (e.g. if the laying area (region) can be divided into a large (countable) number of sections and covering it with a network of lines passing through each of them).

Methods for solving problems of integer linear programming such as placement problems are described quite fully. At the same time, in practically important cases, there are often needs to take into account among the constraints of nonlinear dependencies and/or the presence of continuous functions among inequalities. Such tasks arise, for example, when trying to jointly set routing tasks, laying transport routes and placing various transport infrastructure objects on them.

This is especially evident in the long-term planning of the creation of large container terminals operating in transport hubs, which will have to become the central points of logistics chains and distribution complexes. The lag in the creation of such container terminal complexes (CTC) leads to significant losses, both in the economic and socio-demographic development of individual regions and the country as a whole.

The general formulation of the CTC placement problem can be formulated in the form of a choice of nodal points on the transport network, taking into account the following basic requirements, which just generate the presence of nonlinearities and continuous functions:

- The total maximum volume of investments in the construction and improvement of the CTC is limited:

$$\sum_{i=1}^{N} f_i\left(\sum_{r}^{R} З_r(n)\right) \leq \sum_{i=1}^{N} И_i(n) < И, \tag{5.20}$$

where f_i—functions, $З_r(n)$—costs of the r-th type $(r = \overline{1,\ R})$(design, construction of complexes, transportation, loading, storage of containers, etc.);

- Minimum CTC power $Q_m^{(i)}$ for the reception and dispatch of containers from the regional points covered by it and adjacent points on the transport network should be:

$$Q_m^{(i)} \geq \sum_{l=1}^{\alpha} \phi_l^{(i)}\left(\sum_{j=1}^{\gamma} \vartheta_j(n)\right) \tag{5.21}$$

where $\phi_l\left(\sum_{j=1}^{\gamma} \vartheta_j(n)\right)$—a function of the volume of import–export of containers for the l-th point bordering on the transport network with the i-th point-node;—the number of such points for the planning period n.

In addition, when developing strategic federal plans, a number of other conditions are accepted. For example, it is assumed that in such major transport hubs as Moscow, St. Petersburg, Novorossiysk, Nakhodka, complex CTC are already operating.

The placement of new CTC should coincide with the administrative center of the region, and the expediency, i.e. the rationality of the CTC placement should be determined, first of all, by economic efficiency and socio-demographic benefits (increasing the number of jobs, attracting the population, etc.), reduced to economic indicators.

Then, introducing a set of possible options for the placement of CTC in the nodes of the transport network in the form of X, such that

$$x_i \in X \text{ and } x_i = \{[l_i] \otimes [y_i]\},\ i = \overline{1, N} \tag{5.22}$$

where $[l_i] = \left(l_1^{(i)}, \ldots, l_i^{(i)}, \ldots, l_N^{(i)}\right) = (0, \ldots, 1, \ldots 0)$—vector of options; $y_i = (0, \ldots, y_i(n), \ldots, 0)$—vector column of ranks of the CTC construction sequence in the i-th point ($y_i(n) = 1, 2, 3, \ldots$) transport network in the first, second or third turn.

It is possible to set the task of rational placement in the *i*-th point in the n period. The rationality of placing the CTC in the i-th point in the n period (queue) can be determined by the magnitude of the effect

$$I_{(n)}^{(i)} = \sum_{j=1}^{\gamma} Q_{ji} \left[\phi_l^{(i)} \left(\sum_{j=1}^{\gamma} \vartheta_j(n) \right) \right]$$
$$\left\{ f_{ij}^{(0)}(3(n)) - f_{ij}^{\text{CTC}}(3(n)) + S_{ij} \left(\sum_{j=1}^{\gamma} \vartheta_j(n) \right) \right\}, \tag{5.23}$$

where: Q_{ij}—planned number of container shipments between points I and j of the network;

$f_{ij}^{(0)}(3(n))$—the total reduced costs of container transportation in the existing version before the construction of the CTC ($n = 0$);

$f_{ij}^{(\text{CTC})}(3(n))$—planned costs for the construction of the CTC ($n = 1, 2$);

$S_{ij}(\vartheta(n))$—planned economic benefits from changes in socio-demographic indicators ($n = 1, 2, 3$).

The task of placement is now formulated as the task of choosing such an option x_i, at which the indicator (5.23) reaches its highest value under the constraints (5.20) and (5.21).

To solve the problems under consideration, human–machine procedures using heuristic methods and algorithms are often successful. This makes it possible to include in the calculation process a number of experimentally obtained characteristics, as well as poorly formalized recommendations that take into account the preferences of the DM. One of these algorithms has the following form (see Fig. 5.3).

To fulfill paragraph 5 of this algorithm, namely, to determine the service area of the CTC proposed for placement, the experimentally obtained radii of rational removals are used, found from the lowest total transportation costs between CTC (see Fig. 5.4).

The application of the algorithm shown in Fig. 5.3, taking into account the data obtained from Fig. 5.4, allowed for the first stage ($n = 1$) of the creation of the CTC network of Russia to select 16 points, where, together with the basic ones in Moscow, St. Petersburg, Novorossiysk and the Eastern CTC, it is advisable to place the CTC in Chelyabinsk, Yekaterinburg, Perm, Samara, Rostov-on-Don, Volgograd, Vologda, Kazan, Orenburg, Krasnoyarsk, Tyumen and Omsk.

At the same time, the total projected volume of container traffic between these points will be approximately 37% of the total planned volume of traffic in Russia.

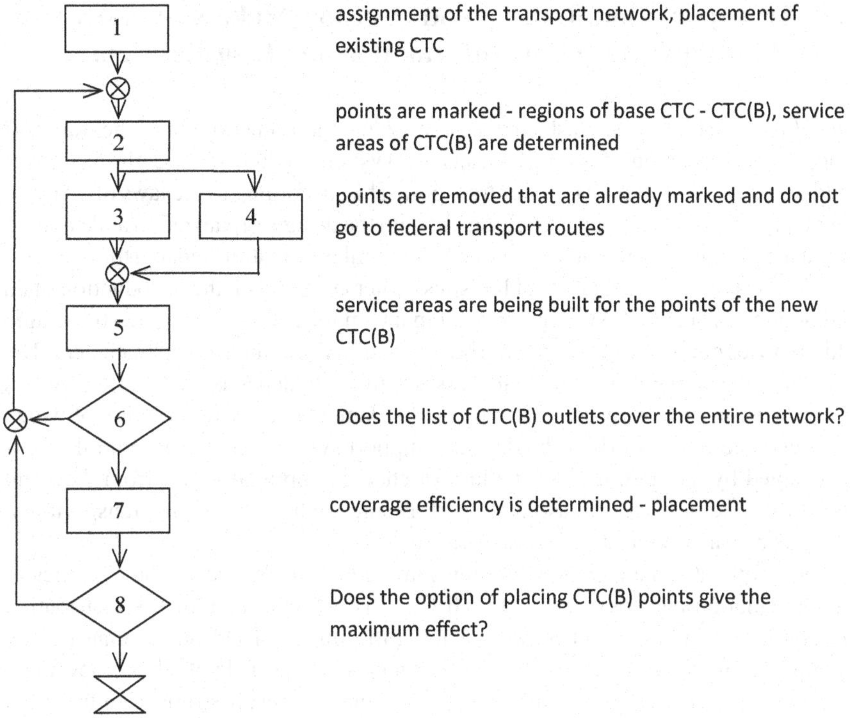

Fig. 5.3 Heuristic algorithm of rational CTC placement: in paragraphs 1, 2, 5, 6—researchers participate; in clause 7, 8—the DM participates; in paragraphs 2, 3, 4, 5, 7, 8—calculations are obtained using a computer

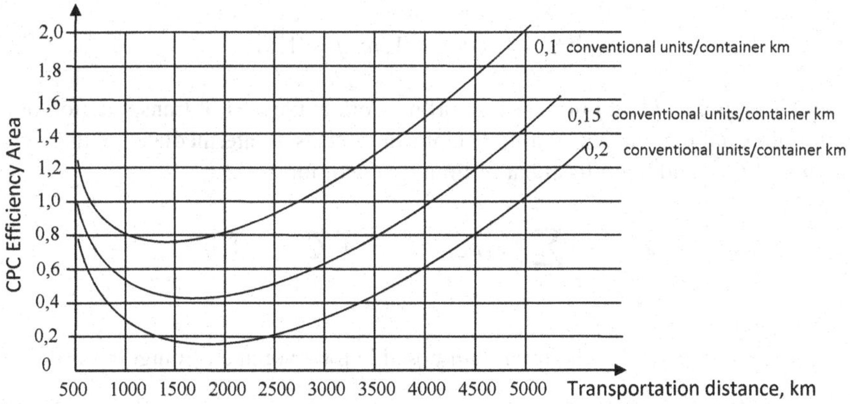

Fig. 5.4 Ratio of total transportation costs between CTC for the cost of transportation from 0.1 to 0.2 euros per container kilometer

5.4 Transportation Routing and Decision-Making Support of the Operator of the Information and Logistics Center

One of the most important tasks in the organization and functioning of the transport space is the formation of a unified transport system, including a freight-passenger conducting network and all types of transport. This, according to the laws of synergy, inevitably leads to innovative development with the design, organization and creation of sound traffic management centers—intersectoral information and logistics centers. The creation of such transport and logistics centers is based on the introduction of an interaction management system for participants in the transport and logistics chain, which carries out its activities with the help of interaction management managers. The success of the activities of such managers significantly depends on the effectiveness of the decision support system. Among the fundamental tasks, the solution of which is the essence of the work of the decision support system, one of the central places is occupied by the routing task, the task of choosing optimal routes from the point of view of one or a set of reasonable performance indicators for the transportation of goods of various types and passengers.

For an operator managing traffic flows in the region covered by the activities of this information and logistics center (ILC), it is essential that such a distribution of vehicles on the transport network known to him is carried out so that a given amount of cargo of various types and passengers is transported with the achievement of the greatest efficiency determined by costs, and as a result, profit from transport activities or other performance indicators—delivery time, reliability, etc. The most general formulation of such a problem will lead to the following mathematical model:

Determine the values of the units of cargo or passengers, that is, the flow of the form (S):

$$x_{ij}^{(S)}\left(S = \overline{1, Q};\ i = \overline{1, n};\ j = \overline{1, m}\right)$$

Between points i and j of the transport network at the cost of transportation of a unit of flow, taking into account the maintenance costs at intermediate points in the amount of $C_{ij}^{(S)}$ and the fulfillment of limiting conditions:

$$a_r^{(S)} \leq \sum_{p \mid P_{(j)}=r} x_{rj} \leq b_r^{(s)}\ s = \overline{1, Q},\ r = \overline{1, n}$$

$a_r^{(S)} = \min \sum_{j \mid P_{(j)}=r}^{m} x_{rj}^{(S)}$—the minimum possible total volume of outgoing flows of the type S from point r.

$b_r^{(S)} = \max \sum_{j \mid P_{(j)}=r}^{m} x_{rj}^{(S)}$—the maximum possible total volume of outgoing flows of the type S from point r.

$$a_i^{(S)} \leq x_{P_{(j)}i}^{(S)} \leq b_i^{(s)}\ s = \overline{1, Q},\ i = \overline{1, ..., r-1, r+1, ..., n}$$

$a_i^{(S)}, b_i^{(S)}$—the minimum and maximum possible flows of the form (S) included in the i-th point. At the same time, the following conditions must be met:

$$\underline{\mu}_{ij} \leq x_{ij}^{(S)} \leq \overline{\mu}_{ij}\ i = \overline{1, n};\ j = \overline{1, m}$$

where $\underline{\mu}_{ij}$ and $\overline{\mu}_{ij}$—lower and upper throughput (x_{ij})branches of the transport network.

$$x_{P_{(j)}i}^{(S)} - \sum_{j \mid P_{(j)}=i} x_{ij}^{(S)} = 0;\ s = \overline{1, Q};\ i = \overline{1, n}.$$

where $P_{(j)}$—the point located before point j in the transport network with the lowest total cost of transportation.

$$J_\Sigma = \sum_{S=1}^{Q}\sum_{i=1}^{n}\sum_{j=1}^{m} x_{ij}^{(S)} c_{ij}^{(S)} \rightarrow \min_{x_{ij}^{(S)}}$$

This linear programming transport problem is comparatively the simplest for the case of $Q = 1$. For the case $Q \geq 2$, the algorithm for solving it becomes significantly more complicated. To solve it, an algorithm based on the reduction of the multithreaded case can be used $Q \geq 2$ to single stream. Indeed, if the transport network with the ILC in point r is represented as a tree-like structure, the root of which is located in point r, and the points of delivery of goods and passengers as the leaves of such an oriented tree, as well as the point $P_{(j)}$ is denoted as a point located before the jth point, excluding point r, then the above statement corresponds to the task of searching for a multi-product ($Q \geq 2$)minimum cost flow. Since the transportation process is associated with the costs of both transportation between neighboring points of the network and through each point, a payment matrix is formed in the expression for the efficiency indicator for each type of cargo $C^{(S)} = \left\{c_{ij}^{(S)}\right\}$, где $c_{ij}^{(S)} = c_{ji}^{(S)}$—the values of non-negative costs for the transportation of a single S-type cargo between points i and j plus the costs at the point of delivery, except for the initial and final points, because they will form constant terms for any route option between points i and j. Task reduction at $Q \geq 2$ to the problem with $Q = 1$ is carried out as follows. Relations of the form are introduced:

$$A_i = \max\left[\sum_{S=1}^{Q} a_i^{(S)}, \underline{\mu}_{P_{(i)},i}\right]$$

$$B_i = \min\left[\sum_{s=1}^{Q} b_i^{(s)}, \overline{\mu}_{P_{ij},i}\right], (i = \overline{1, \ldots, r-1, r+1, \ldots, n})$$

$$A_r = \sum_{S=1}^{Q} a_r^{(S)} \quad B_r = \sum_{S=1}^{Q} b_r^{(S)}$$

Then the problem with Q flows of transported goods of various types. ($Q \geq 2$) boils down to a task of the form:

$$A_i \leq y_i \leq B_i; \; y_i \geq 0; \; i = \overline{1, n}$$

$y_i = \sum_{j|P_{(j)}=i} y_j; \; i = \overline{1, n}$ except for the end points $j = \overline{1, m}$

$$\tilde{J}_\Sigma = \sum_{i=1,\ldots,r-1,r+1,\ldots,n} y_i \tilde{c}_i$$

where $y_r = \sum_{i|P_{(i)}=r} x_{ri}; \; y_i = x_{P_{(i)}i}; \; \tilde{c}_i = c_{P_{(i)}i}; \;\; i = \overline{1, \ldots, r-1, r+1, \ldots, n}.$

With this limitation in the task ($Q \geq 2$) form a joint system only if $A_i \leq B_i$; ($i = \overline{1, n}$).

The solution of the problem here is relatively simple using solutions of linear transport programming or streaming algorithms, as well as algorithms based on the dynamic programming method. Using the given ratios and other transformations, the components of transportation $Q \geq 2$ using container or batch processing in order to minimize Q, it is possible to build a knowledge base of the automated decision support system of the ILC transport management manager.

Chapter 6
Modeling of Decision-Making Processes in Transport Management

6.1 Taking into Account the Human Factor and DMP in Transport Management

For the rational construction of transport process management, it is necessary to take into account the possibilities of DM, while separating the limitations that are characteristic of a person in general (HF) from the moments associated with negligence, lack of responsibility and readiness for DMP, lack of education or a painful condition, usually attributed to the influence of a personal factor (PF).

Often, the manifestations of PF also include those favorite techniques and procedures that are used specifically by this DM in the DMP in the management of transport processes (MTP). They, along with the listed negative traits, characterize personality traits, the presence of which in traffic accidents acts as the main or concomitant cause—a personal factor.

The results of long-term observations of the activities of the DM in such transport systems as air traffic control, water transport management, railway traffic management, etc., convincingly show that the DM has a number of disadvantages and advantages in DMP. For example, in solving the problem of choosing an option in conditions of uncertainty, the DM has undeniable advantages, but it is quite difficult to investigate the logic of the DM, to find in it the absence of errors and contradictions, and there is no exhaustive data here yet. This is what determines such a large number of DMP methods, the authors of which proceeded from each of their own models of DMP behavior.

At the same time, if we do not take into account the influence of SF and LV in the blocks of the typical TPM structure (Fig. 3.1), then the most constructive methods can be considered to be of the axiomatic type.

The accounting of HF and PF in the activities of various types of DM can be traced, for example, through reviews and monographs related to aviation, railway, and other transport systems.

Usually, for a DM of the type 212, 211, 222, 221, according to (Fig. 3.2), its capabilities are evaluated on the 4th indicators:

© The Author(s), under exclusive license to Springer Nature Singapore Pte Ltd. 2023

G. A. Kryzhanovsky et al., *Modeling of Transportation Aviation Processes*, Springer Aerospace Technology, https://doi.org/10.1007/978-981-19-7607-0_6

- formation of the decisive rule;
- comparison of DMP with objective data;
- transitivity of DMP:

$$\begin{aligned} &A > B + B > C \rightarrow A > C \\ &A > B + B = C \rightarrow A > C \\ &A = B + B = C \rightarrow A = C \end{aligned} \tag{6.1}$$

- consistency and stability DMP:

$$\{2.1.1(t - \tau) = 2.1.1(\tau) = 2.1.1(t + \tau)\}$$

Thus, when solving the problem of choosing the best option and ranking or classifying options for many indicators of this type of DM, non-transitivity, inconsistency, and the desire to apply primitive decision rules based on simplification of the task, often distorting their policy, were allowed. Moreover, the analysis of various methods of assigning weight coefficients shows that there are no correct methods for justifying their choice by a person yet—the same effects of PF and HF are traced in the DMP.

At the same time, taking into account (parrying) the influence of PF can be carried out mainly through psychophysiological selection and vocational training (training) DM or, finally, the development of automated support, while the reason for the Black Sea is limited human capabilities. Some of the most characteristic limitations for the DM operator of the form 111, 112, 121, 122 are quite widely known.

In addition, for LPR of any type are known:

- limitations on the capacity of short-term memory, which is 7 ± 2 blocks of structural units of information;
- an almost unlimited amount of long-term memory and its participation in DMP in the conditions of typical situations when "learning" DM by trial and error;
- single-channel with sequential processing of a small amount of information, but with an associative system for its search;
- adaptability, expressed either in adapting to the type of DMP tasks that exceed the complexity of those previously known to him, in case it is impossible to change the circumstances (situation, conditions), or in simplifying the DMP task, adapting it to the level of its capabilities;
- a significant influence of motivation on such certain parameters of the DMP as its speed, generation of alternatives and depth of search for solutions, but not overcoming the limits of the limitations of the HF;
- a significant influence of the dimension—complexity of the DMP task as a consequence of the limited capabilities of the DM in terms of short-term memory (Fig. 6.1), i.e. sensitivity to the dimension—complexity of the DMP.

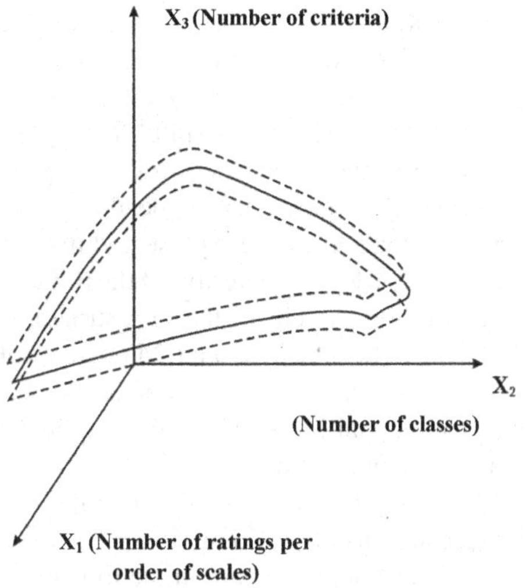

Fig. 6.1 Group capabilities of the DM in solving the problem of object classification

Usually, the following parameters of DMP tasks are taken into account: the number of performance indicators, the nature of the scales (discrete, continuity), the number of quality gradations with discrete scales, the number of alternative options, the nature of their assessments on the scales of indicators (discrete, continuous, approximate, accurate) (Fig. 6.1).

So it is experimentally confirmed that for values $S(x_1, x_2, x_3) \geq 7.2$ the behavior of the DM becomes non-transitive, often contradictory and error-prone. At the same time, these phenomena can be excluded (parried) in several ways, and the presence of the Black Sea Fleet should still be taken into account on each of them. Such ways are known and, basically, it is either the development of a holistic image in the DM with the help of training and training, when instead of taking into account a separate final situation S, the DM "thinks", represents the general situation, for example, the image of the DTS, automation of the DM with the help of the entire arsenal of science, the use of the institute of research consultants, finally, borrowing successful variants of DMP in the TSM by using the results of solving the inverse optimization problem of DMP.

Each of the listed directions and some others will find their reflection below in the degree of depth and breadth of coverage available to the authors: we are far from thinking of giving an exhaustive exposition of the science of transport process management and such a fundamental problem as DMP at TPM. Here, a more modest, more "grounded" approach has been adopted, which consists in understanding everything that has already been developed in the field of TPM and as close as possible to modeling, and, where this is permissible, to optimizing the DMP in TPM. For example, clarifying the model of the DMP, it is also worth noting that the efficiency of the DMP is significantly affected by the DMP time. If the general idea of the formation and development of a solution, which can be obtained from (Figs. 3.6 and

3.7), is supplemented with another one—some simplified algorithm for the development of the SFA and the development of the SFA, as well as the construction of a holistic image (Fig. 6.2), then it becomes clear the need for the execution time of each block in such, even a simplified algorithm.

However, the dynamics of transport processes leads to significant changes in the consequences of decisions depending on the time of issuing the command—SFA. This dependence can be represented in the form of a curve shown in (Fig. 6.3). It is clear that the degree (severity) of the influence of the FA generation time for different levels of the hierarchy of the SFA structure is different and is most significantly felt at the lower level, where the DM operator of the ergatic transport system sometimes has to act in fractions of a second. In this case, the main role is played, of course, by his professional level—the skills acquired during training, training and experience in the development of SFA.

Assuming the indifference of time in relation to the quality of the SFA in the representation of the DM on (Fig. 6.3), five characteristic zones can be distinguished. The first is where any gain in the time of receiving and processing information and

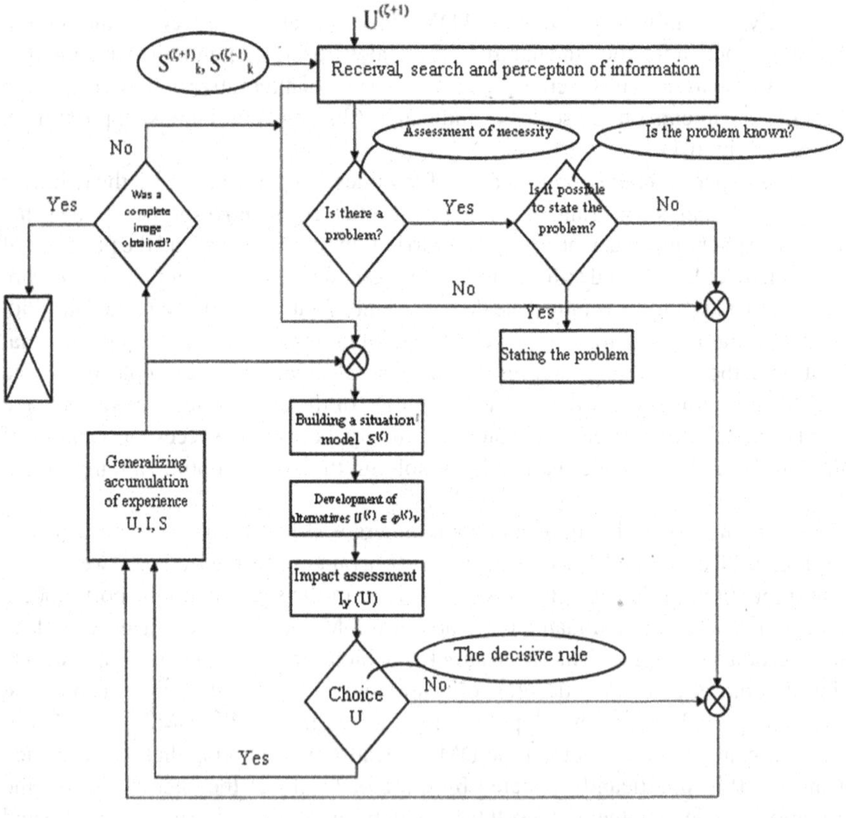

Fig. 6.2 Simplified algorithm for the selection of SFA and the construction of a holistic image

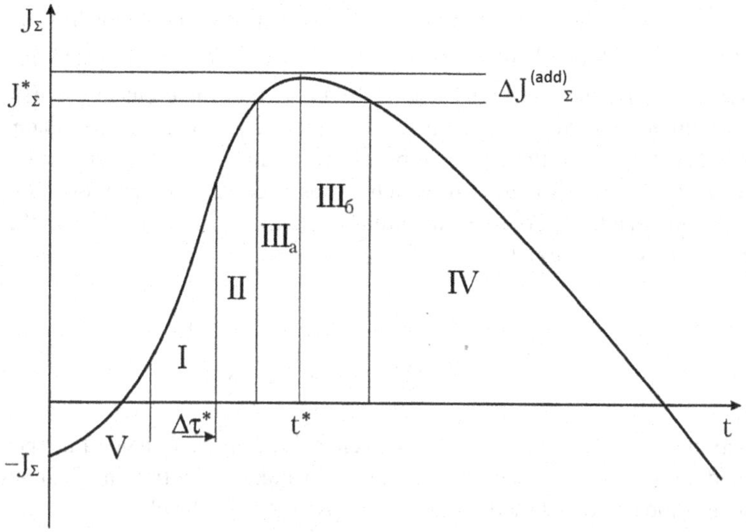

Fig. 6.3 Model of the dependence of the effectiveness of the solution on the duration of the DMP

the development of the SFA gives a significant effect in the increase in the value of the indicator. The second zone is characterized by the fact that significant information has already been received and delays in the DMP for obtaining additional information or developing new alternatives to the SFA do not lead to a sharp increase in the value of the indicator. The third zone, where the quality of DMP practically does not depend on the time of data search and the development of alternatives, here the change in the value of the efficiency indicator in relation to its maximum value does not exceed some small value $\Delta\ J_{\Sigma}^{(\mathrm{add})}$. In this zone, as a rule, the characteristics of the activities of the most experienced, successful and professionally trained DM are laid down. Finally, the fourth zone, where significant delays in the DMP lead to a significant decrease in the efficiency indicator and often makes the SFA practically meaningless, as well as attempts to act with a lack of information and the development of an alternative—when $t \leq \Delta\tau^*$ in the zone V.

It is clear that the model (Fig. 6.3) gives only a general idea of the role of time in the DM in the form of a typical characteristic curve, specific numerical values on which can be found experimentally for each of the variants of the situation, for example, during training of a particular DM. At the same time, it is necessary, of course, to take into account the psychophysiological state of the DM. In this case, it is possible to set the amount of the "fee" with which he pays for the implementation of the DMP in this situation. This is necessary for a variety of reasons—both to determine the degree of its stability and performance in the future, and to determine its classiness and other characteristics.

Thus, by constructing a probability distribution function for choosing certain solutions (including: erroneous omissions, negligence, etc.), which characterize certain motives—personal preferences, performance indicators of this $\mathit{ЛПР}_k$, correlated with objective assessments of the situation and instructions-commands from a higher level $U^{(\Im+1)}$, as well as plans $y^{(\Im)} - W_k^{(\Im)}\left[p_{k\ell}^{(\Im)}(U), I_{kv}^{(\Im)}\left(U^{(\Im+1)}\right), S(x)\right]$, and also taking into account the characteristics of the DMP time inherent in this DM in a particular situation with its preferences—$\tau_k^{(\Im)}\left(I_{rv}^{(\Im)}, S\right)$, and having supplemented all this with estimates of the vector of psychophysiological costs, also attributed to an objectively assessed situation in which it was necessary to carry out SFA, it is possible to obtain a fairly complete quantitative characteristic of the DM in the form of a certain vector-functional:

$$V_k^{(\Im)}(S) = V_k^{(\Im)}\{W_k^{(\Im)}\left[p_{kl}, I_{kv}^{(\Im)}, S(x)\right],\\ \tau_k^{(\Im)}\left(I_{kv}^{(\Im)}, S\right), \Phi_k^{(\Im)}(U, S)\} \tag{6.2}$$

Functional (6.2) serves essentially as a generalized characteristic of the personal factor and, in its physical essence, just like the Lyapunov function in dynamics, can determine some achievable k-th "energy" in space $V^{(\Im)}(S)$ level.

Functional $V_k^{(\Im)}$ characterizes the activity DM_k, for example, the way k-th $\text{DM}_k^{(S)}$ transforms the circumstances of a given situation $S_r^{(\Im)}$ alternative possibilities $U_{kl}^{(\Im)}$ with efficiencies $I_{kv}^{(\Im)}$ and instructions $\text{DM}_k^{(\Im+1)} \to U^{(\Im+1)}, y^{(\Im+1)}$ into an individual expected value with a delay $\tau_k^{(\Im)}$ and psychophysiological costs $\Phi_k^{(\Im)}$. To do this, the functionality (6.2) must be presented as a convolution:

$$V_k^{(\Im)}(S) = \sum_{\nu k=1}^{\mu k} \omega_{\nu k}^{(\Im)} \sum_{i=1}^{n_\Im} \sum_{\ell=1}^{m_\Im} p_{k\ell}^{(\Im)} I_{\nu k}^{(\Im)} \Psi_{ik}^{(\Im)}(x) \tag{6.3}$$

where $\Psi_{ik}^{(\Im)}(x) \in \varphi_\Psi^{(\Im)}$—the set of individual values of each 1st result for a given k-th DM_k on $\Im$-M the level of the TSM hierarchy in the situation $S_r^{(\Im)}$, $\omega_{vk}^{(\Im)}$—the weight coefficients of the ν-th indicator on $\Im$ the level of the TSM hierarchy for this DM_k:

$$\sum_{\nu=1}^{\mu_\Im} \omega_{\nu k}^{(\Im)} - 1 = 0; \quad \omega_{\nu k}^{(\Im)} \geq 0 \tag{6.4}$$

Using a convolution of the form (6.3) allows you to evaluate changes in individual value DM_k depending on the information received, which makes it possible, in turn, to introduce some scale of evaluation of the information itself. This approach to the evaluation of information as a vector quantity through the evaluation of changes in the vector value of personal characteristics $V_k^{(\Im)}$ allows you to evaluate the information component that characterizes changes in the state of the control object, the information component that carries instructions for evaluating events, i.e. changing the values "from above". $I_{kv}^{(\Im)}$, and finally, information that changes motivations—individual values DM_k. At the same time, the estimates obtained are not determined by the number of transmitted signals, but by the reaction DM_k—change $V_k^{(\Im)}$. It is clear that, depending on the essence of the tasks set for the study and optimization of DMP in the TSM, the type of convolution of the vector-functional $V^{(\Im)}(S)$ can be different and must meet (correspond to) this essence.

A special role is played by the duration of the DMP in the TPM in ensuring transport safety. Examples of dangerous approaches and collisions of aircraft in air transport speak most eloquently about this. An essential role in ensuring safety here is played by the ATC system and especially by the radar control dispatcher. The most important characteristic of his activity is his ability to predict the DAS.

Based on the results obtained, graphical dependences of the forecast error on the forecasting period in the situation are constructed S_1 (Fig. 3.1).

Analyzing the results of the experiment, we can draw the following first conclusions (assumptions):

- prediction error $\overline{\delta}_D^{(S,\alpha)}$ period in the situation S_1 of the DAS state increases with the increase in the duration of forecasting for both groups of dispatchers;
- the law of changing the forecast error in a situation S_1 the state of the DAS from the duration, close to the parabola (in the study area $\Delta t = 1 \div 6$ min) for both groups and all cases of using information support options;
- for groups of dispatchers with a loss of skills in providing forecasting, DAS forecast errors are greater than for a group of current airport controllers.

Pulkovo; the biggest difference $\Delta_\sigma^{(\alpha)} = \delta_D^{(1\alpha)} - \delta_D^{(2\alpha)} = 1$ km when forecasting for the period $\Delta t = 4$ min. It should be noted that when $\Delta t > 4$ min there is a reduction in the difference in estimates b ^ between dispatchers and in the vicinity of the point $\Delta t = 6$ min these estimates are almost equal. This is due to the fact that the magnitude of the error $\overline{\delta}_D^{(S,\alpha)}$ related to the qualification factor (in case of loss of skills), compensated by a general error $\overline{\delta}_D^{(S,\alpha)}(\infty)$, related to accumulation during $\Delta t \to 7$ min errors caused by random interference ξ_1 and $\xi_2[M(\xi_1) = 0$ and $M(\xi_2) \neq 0]$, necessarily existing in the real state of the DVO with any variant of information support, the nature of which is related to the accuracy of aircraft navigation, wind gusts, etc.;

- spread of values $\overline{\sigma}_{\delta D}^{(2,\alpha)}$ for a 1st class dispatcher with a loss of skill less than that of a 1st class dispatcher at Pulkovo Airport. This is due to standardization in the simulation of flight modes of aircraft on the simulator, which facilitates the prediction of their movement.

Having empirical material on evaluation $\overline{\delta}_D^{(S,\alpha)}$ for the dispatcher of the 1st class, you can build a mathematical model—$\overline{\delta}_D^{(1,\alpha)} = f_M^\alpha(\Delta t)$ in the form of a second-order equation:

$$\overline{\delta}_D^{(M,\alpha)} = a_0 + a_1 \Delta t + a_2 (\Delta t)^2.$$

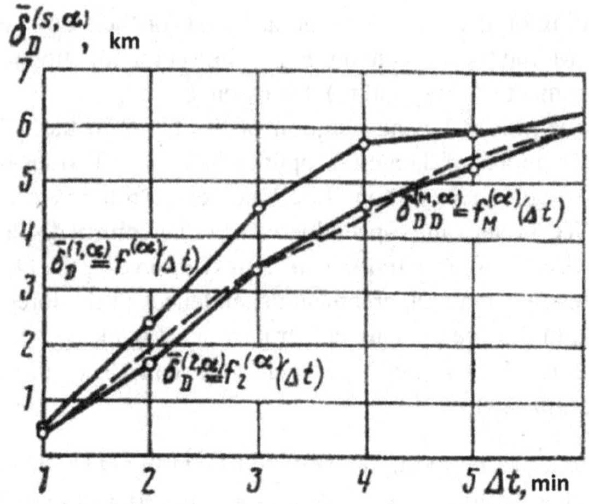

Fig. 6.4 Empirical dependencies $\overline{\delta}_D^{S,\alpha} = f_S^{(\alpha)}(\Delta t)$ for dispatchers of two groups of the 1st class—Pulkovo Airport and students at the University of GA—from the duration of the forecast: $\delta_D^{1,\alpha} = f_1^{\alpha}(\Delta t)$—the function of the DAS prediction error by the range coordinate for the dispatcher of the 1st class of the University of GA; $\delta_D^{2,\alpha} = f_2^{\alpha}(\Delta t)$—the function of the DAS prediction error by the range coordinate for the dispatcher of the 1st class of Pulkovo Airport; $\delta_{DD}^{M,\alpha} = f_M^{(\alpha)}(\Delta t)$—a mathematical model describing the dependence of the forecast error on the duration of forecasting for the dispatcher of the 1st class

Using the procedure of the least squares method, it is possible to write the equation of the model describing the dependence $\overline{\delta}_D^{(1,\alpha)} = f_M^{(\alpha)}(\Delta t)$ д S_1 condition DAS, $\overline{\delta}_D^{(M,\alpha)} = -1.52 + 2.045\Delta t - 0.132(\Delta t)^2$.

Graphically, the mathematical model is shown in Fig. 6.4.

The constructed mathematical model can be used in the study of other characteristics of the DMP by ATC dispatchers and in assessing the qualifications of the dispatching staff of the civil aviation traffic service. Mathematical models describing dependence $\overline{\delta}_D^{(S,\alpha)} = f_S^{(\alpha)}(\Delta t)$ for dispatchers of the S-th professional level ($S = 1$)for various types (modifications) of information support are shown in Fig. 6.5. When $\alpha = 1, 2, 3$ где $\alpha = 1$ corresponds to the case of AC ATC, $\alpha = 2$—the use of radar and VRL "Sign", and $\alpha = 3$—application of radar.

The considered parameters of the DMP and their models as a whole represent a certain part of the objective characteristics of the influence of the quality of information support and the qualification level of the dispatching staff. The value of these parameters seems to consist in the fact that they can be experimentally obtained each time during the study of various information support options and can show the trend of their changes in the activities of ATC dispatchers. And if so, then with their help it is necessary to evaluate and improve the information support of the decision-making process. Consequently, the quality of the decision-making process at the direct ATC can be improved by improving information support and improving the professional level of the dispatcher.

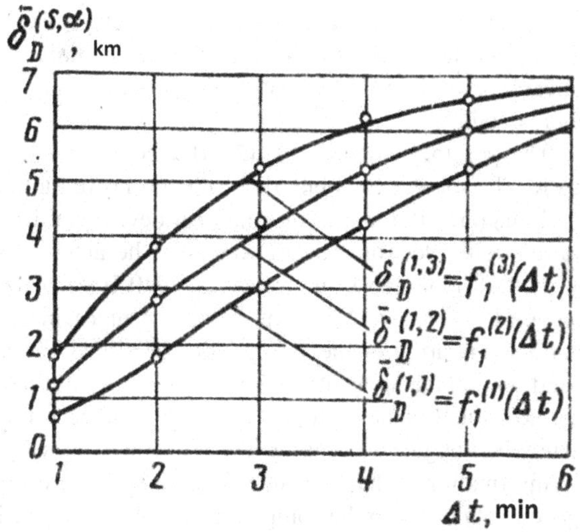

Fig. 6.5 Dependence $\bar{\delta}_D^{S,\alpha} = f_S^{\alpha}(\Delta t)$ in a typical DAS situation for dispatchers of the 1st class $S = 1$ with various options $\alpha = 1, 2, 3$ from the time of updating the goal marker

However, as the practice of ATC shows, the traditional options for the information support of the DMP of the dispatcher of the direct ATC at the level of the second- and third-generation ATC do not provide a complete, sufficient guarantee of its error-free operation. The new generation of the ATC control system should contain not only information support for the development of the DAS, but also elements of support for the DMP by the ATC dispatcher. The most well-known options for such support include collision prevention systems (CPS). Aircraft collision prevention is a complex problem, the solution of which is caused by multiple duplication of preventive measures, starting with the separation of air routes and corridors, planning processes and ending with the development and implementation of automated support tools for the direct ATC dispatcher and/or crew warning. Such CPS are developed in the USA. In Japan and other countries have been researched for more than 40 years. In our country, the development of the CPS covers various areas, including the creation of airborne (ACPS), ground-based (GCPS) and integrated (ICPS) systems. Recently, more and more attention has been attracted by a variant of the CPS based on the use of a secondary radar system with a discrete address request—the so-called VRL mode system (SVRL-S). The solution to the problem of collision prevention based on the SVRL-S is implemented by using an individual address request, a high-speed data transmission line (HSDTL), monopulse bearing in the address system, which significantly increases the accuracy of measuring the coordinates of targets and increases the reliability of DAS surveillance. The advantage of such a variant of the ICPS is based on more complete information about the relative movement of aircraft than in the ACPS which allows solving conflict situations between them not only by vertical, but also by horizontal maneuvering, coordinating evasion decisions with the general DAS in this ATC zone or, most importantly in the case under consideration, to implement the idea of integrating a number of the simplest ATC circuits, covered by this SVRL-Ss ICPS on its basis. Since such a ICPS is one of the most promising subsystems of the ASPI of direct ATC, which provides support for the radar control dispatcher during conflict resolution, and, in addition, serves as an example of the integration of the SVRL-S within the ATC, it is advisable to consider its features in more detail.

The possible structure of the ATC control system complex together with the SVRL-To solve the problem of collision prevention can be presented in the form of a diagram (Fig. 6.6). Antenna drill device can be made in two versions. In the variant with electronic beam scanning, the antenna is a cylindrical active phased array with radiation in the range of 1030/1090 MHz and an annular array in the range of 740 MHz. A variant of the antenna with mechanical beam rotation is a flat antenna array of the 1030/1090 MHz band and a linear row of the 740 MHz band. In both versions of antennas, three types of directional patterns are formed in the horizontal plane: total, difference and interference suppression. Measuring the azimuth of targets with monopulse processing allows you to get an accuracy of 5–8′.

Information exchange with other elements of the complex within the ATC control system, mainly with the computing complex, is carried out through high-speed data transmission equipment. SVRL-S will prevail with increased accuracy of coordinate determination, noise immunity against internal interference, high data transmission rate over the "ground—board—ground" LPD.

The main difference between this element of the ICPS and the usual VRL system is the new principle of target selection – instead of the spatial method, an address

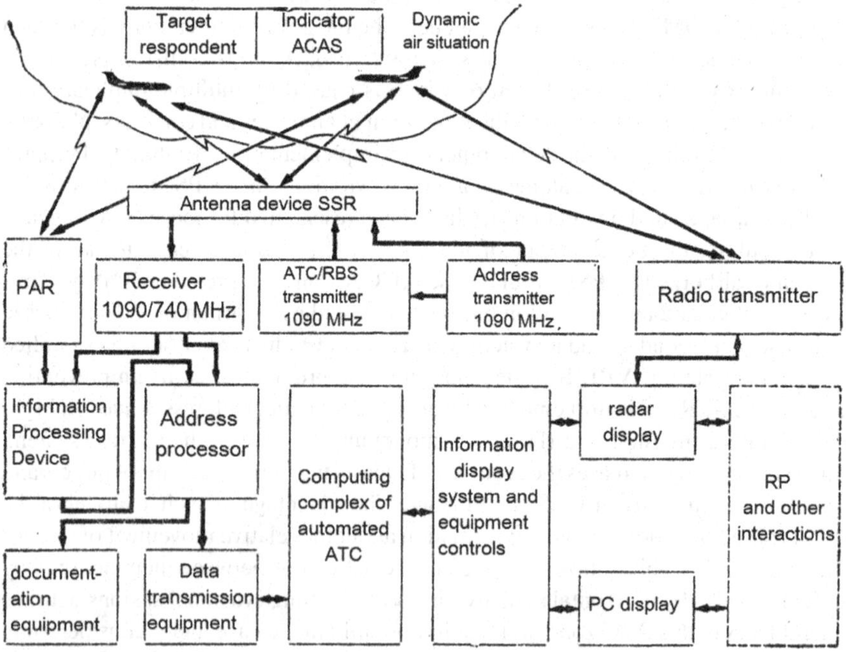

Fig. 6.6 Structural diagram of the CPS based on the ATC drill-S

query is applied, when each aircraft in this DAS is assigned its own unchanged 24-bit address code. Integration within the framework of the AC ATC is carried out by combining DRILLS-S with existing drills and including the functions of the CPS in the computing complex of the AC ATC. The combination is carried out using the same radio frequencies on request and response and the same formats of request signals. In the general request mode, an additional pulse P4 with a duration of 0.8 microseconds is activated, following the pulse P3 after 2 microseconds, to which (P4) a response in the form of an individual identification code follows. The address request begins with two pulses P1 and P2 with corresponding pauses (Fig. 6.7).

Relative phase modulation is used for data transmission, the speed of such an LPD is 4 Mbit/s. The response signal consists of a response key with a duration of 8 microseconds, including four pulses, and an information block containing 56 or 112 binary digits using time pulse modulation (Fig. 6.8).

The maximum throughput of the address system reaches 10 responders in the beam of the radiation pattern. Address responders are filtered in the general call and address request modes. For temporary compression, as well as to prevent overlapping

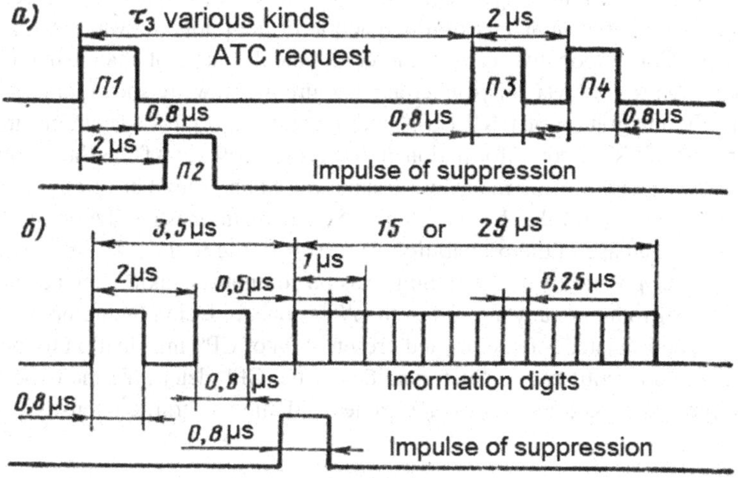

Fig. 6.7 General (a) and address (b) requests

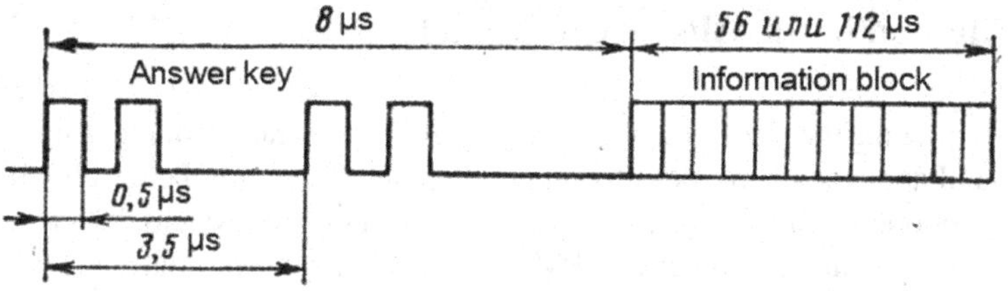

Fig. 6.8 Address response

of request and response signals in the address system, requests are ordered to several aircraft in the beam of the radiation pattern. The main characteristics of such and as an element of promising information support for direct ATC can be given in the form of Table 6.2. Part of the data was obtained by statistical testing and semi-natural modeling on a complex including flight and control simulators and basically coincides with the data given for CPS.

The information of such a CPS is the support of DMP for two types of DM at once: the ATC dispatcher and the pilot (crew). At the same time, the crew is provided with two types of information: advisory and command. The advisory information allows the crew to increase the probability of visual detection of potentially dangerous aircraft, simplify the assessment of the degree of danger on their part, and choose a safe and effective evasion maneuver. This allows collision avoidance with minimal change of course, which in turn leads to the least disruption of air traffic plans. This effect is important because it reduces the likelihood of new conflict situations and eliminates the need for the crew to abruptly change course at the very last moment in stress conditions. The stages of warning, i.e. the support of the DMP of the crews of conflicting aircraft and ATC controllers from the CPS, can be represented as a series of time intervals depending on the characteristics of the CPS (see Table 6.1) and on the data characterizing the reaction of the pilots (crews) LA and ATC dispatchers. Conventionally, such time intervals and the curve of change in the value of information for the ATC dispatcher and the aircraft crew are shown in Fig. 6.9.

Thus, the integration of DRILLS-S within the ATC control system ensures the development of PR support information in such critical cases at ATC as the occurrence of a conflict situation. This makes it possible to obtain a number of new qualities for information support that have such a CPS. The main ones are the possibility of a significant increase in the probability of safe air traffic in the ATC zone with the ATC due to the possibility of developing evasion commands, taking into account the entire DAS surrounding any aircraft in the ATC zone, reducing the number of false alarms and coordinated interconnected presentation of CPS information to the ATC dispatcher and aircraft crews. Such qualities of the CPS determine the expediency of its implementation primarily in ATC zones with high air traffic intensity.

6.2 The Decision-Making Process in the TSM in the Presence of Models. Analytical Hierarchy Method in the Absence of Models

When automating the USP processes at any level of the hierarchy of the management system and at any stage of its functioning, the main issues are the development of methods for optimizing interaction $DM_k^{(\Im)}$ with an automated system, which today is most often a personal computer or $APM^{(\Im)}$—the automated workplace is for $\Im$ level of the ATC. Among such methods, the central place is occupied by the methods of optimizing the DMP PC and the implementation of the dialog mode of operation

Table 6.1 Data of the averaged statistical analysis of the forecasting error of the DAS

Type of AT controller	Period of forecast Δt, minutes											
	$\Delta t = 1$		$\Delta t = 2$		$\Delta t = 3$		$\Delta t = 4$		$\Delta t = 5$		$\Delta t = 6$	
	Evaluation of the random variable											
	$\overline{\delta}_D$	$\overline{\sigma}_D$	$\overline{\delta}_D$	$\overline{\sigma}_D$	$\overline{\delta}_D$	$\overline{\sigma}_D$	$\overline{\delta}_D$	$\overline{\sigma}_D$	$\overline{\delta}_D$	$\overline{\sigma}_D$	$\overline{\delta}_D$	$\overline{\sigma}_D$
Without the loss of skills: 1st class ATC at Pulkovo Airport	0.5	0.48	1.8	1.44	3.5	2.08	4.7	2.7	5.3	3.5	6.0	4.7
With the loss of skills: 1st class ATC; studying at the Academy of Civil Aviation	0.5	0.48	2.2	0.9	4.2	1.41	5.7	2.1	5.9	2.55	6.2	3.8

Table 6.2 Characteristics of the S mode radar

Characteristic name	Unit of measurement	Symbol	Value
Mode *S* range for the aerodrome/route area	km	D_{PA}/D_{Tp}	160/400
Shortest range	»		1/1
Elevation field of view	degree	θ	0.5–47
Coordinate measurement accuracy: by range » azimuth	M Hour Angle per minute	δ_D δ_θ	50 5–8
Rate of view for radar	Revolution/minute	Ω_{ob}	10–20/5–10
Request signal frequency	MHz	f_a	1030 ± 0.01
Response frequency	»	f_o	1090 ± 1
Beam width	degree	θ_a	≥3.5
Resolution	Number of aircraft	N_A	Any two independent aircraft
The number of processed response signals—targets in the radar	Number of aircraft	N_M	100/300
Mean time between failures		T_{OT}	4000
Probability of obtaining undistorted information	–	$P_I{}^{DM}$	≤0.98
Probability of visual detection of a conflicting aircraft by the pilot during an alarm	–	$P_p{}^{SPS}$	~0.65
The probability of critical approach during the actions of the pilot after visual detection of a conflicting aircraft	–	$P_{KVP}{}^{SPS}$	~0.05
The probability of waiting for a command from the SPS by the pilot after visual detection of a conflicting aircraft	–	$P_p{}^{SPS}$	~0.5
The probability of critical approach during the actions of the pilot after the command to maneuver from the SPS	–	$P_{KMP}{}^{SPS}$	~0.05

$DM_k^{(\Im)}$ from a PC or in an ARM, which are carriers of arrays of information organized into intelligent databases with built-in systems for monitoring, processing and accumulation of information.

At the same time, in the control structure, the simplified scheme of which is shown in (Figs. 3.8 and 3.9), now, instead of one block $DM_k^{(\Im)}$, you can imagine a detailed structure that combines the actual $DM_k^{(\Im)}$ with $APM^{(\Im)}$ or a PC, or $DM_k^{(\Im)}$ working with a PC or an entire computing complex through an "intermediary"—a group of research consultants (Fig. 6.10). Examples of schemes that have already

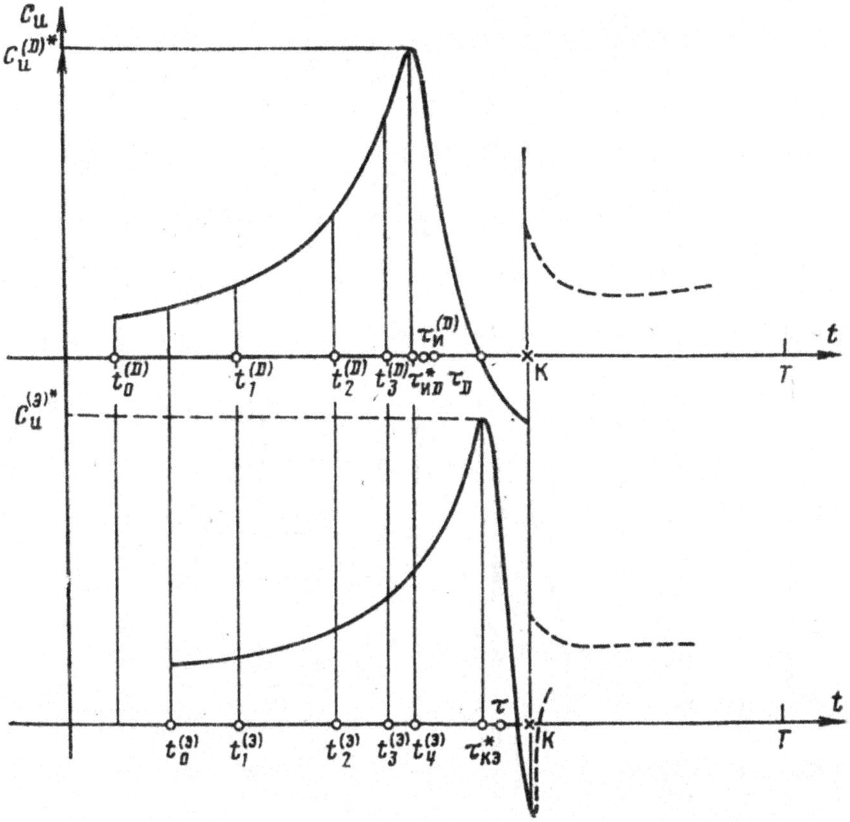

Fig. 6.9 Change in the value of information and time intervals of ICPS signals for the ATC dispatcher and the aircraft crew: *K*—estimated time of the crash (collision); $t_0^{(D)}$—the beginning of the OS threat; $t_0^{(э)}$—the beginning of a collision threat; $t_1^{(D)}$—the CPS signal about the occurrence of a OS; $t_1^{(э)}$—CPS information on the presence of a conflicting aircraft; $t_2^{(D)}$—CPS signal about the threat of collision; $t_2^{(э)}$—the time of the greatest value of information for the dispatcher; $t_2^{(э)}$—the time of greatest value for the crew; $\tau_{Кэ}^{*}$—estimation of the time limit for the reaction of the dispatcher, crew, aircraft; τ_D^{*}—the message of the CPS about the commands transmitted to the crew to ban the maneuver; $t_3^{(D)}$—the CPS command to ban the maneuver; $t_3^{(э)}$—the message of the CPS on the transfer of commands to the crew to perform an evasion maneuver; $t_4^{(э)}$—the command of the CPS to perform an evasion maneuver

been implemented in railway, aviation and maritime transport, and which contain the main points of work of groups of research consultants with automated control systems—prototypes of an intelligent system (IS)—in dialogue mode can be given (Fig. 6.11).

In such schemes, it is possible to conditionally distinguish functions performed by a person (a group of research consultants) and functions that can be performed on an IS. At the same time, the work of the general algorithm implementing the

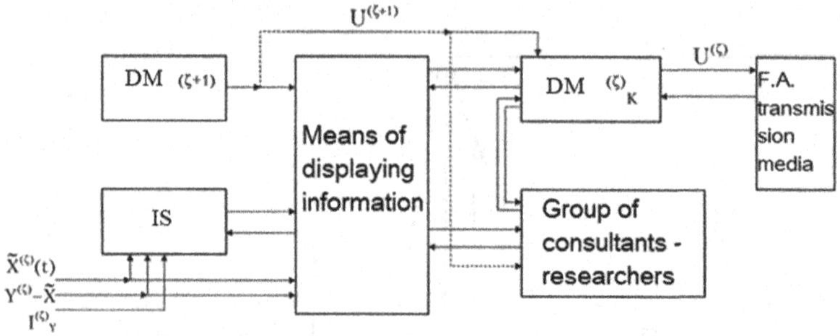

Fig. 6.10 Scheme of interaction of the DM with a group of consultants

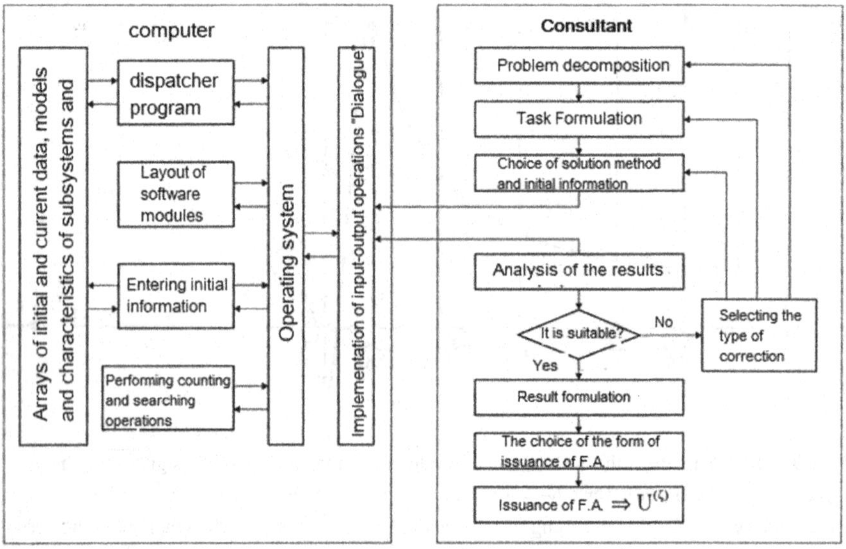

Fig. 6.11 Functional diagram of the implementation of the dialog mode of operation DM^{ζ} and/or a research consultant and a computer (IS)

dialog mode began with the division (decomposition) of the problem into a number of tasks and the formulation of one or more of them (within the framework of the general purpose of the study). Next, the researcher (or their group) selects a suitable mathematical model of the conditions of the problem and the method of its solution and possible outcomes—the values of the vector of the state of the control object—and the available control tools—alternatives to the SFA for $DM_k^{(\Im)}$.

These operations have not yet been automated in real TPM of Russia and were related to the creative activity of the DM or consultants based on their experience and intuition, supported by knowledge of applied issues of the theory of operation research, specific characteristics of the investigated DMP and means of their implementation.

Since it was believed that the number of different types of process models in TPM is not so large, and the number of state values for each level depends on the nature of the problem being solved, it was assumed that in some cases it is possible to have a special array of variants of typical states on a machine medium in the form of, for example, Markov models with discrete and continuous time or models another convenient view. The initial data for the selected model and the corresponding variant of the state of the ergatic transport system and the available controls are selected from arrays of characteristics of their states and efficiency, which are stored on automated control system machine media, periodically adjusted and updated.

DM or consultants, in dialogue with the automated control system, enter data into ready-made programs, receive and analyze the results of solving problems. The degree of their connection with the ACS in this case may be different, for example, three of the most characteristic cases can be distinguished:

- The DM or the research consultant prepares and enters the initial data himself in accordance with the requirements of the program, while observing the rules of data ordering, their location and formats.
- All preparation and data entry work are done automatically. The possibilities of varying the source data are limited by the composition of control programs and a fixed choice of data.
- The prepared data are provided to the DM or to a research consultant who performs their variation and input, while systematization and revision of data are possible.

Other data options are also possible.

One of the most common options for the implementation of such a dialogue can be called the implementation of a solution to the problem of multi-criteria choice at DMP, typical for all levels of TPM. At the same time, an approach is used to identify the preferences of the DM simultaneously with the definition-prediction (miscalculation) of a set of acceptable solutions.

However, the problem of rational distribution of functions between the DM and/or the consultant researcher, on the one hand, and the automated control system, on the other, is still not fully solved in this case. For example, the tasks of finding a so-called satisfactory solution, the assignment task, and many others that are often encountered in TPM, are reduced to the following optimization task with many performance indicators:

Find additional corrective controls $\delta u_l^{(\Im)}$ and parameters $\delta p_q^{(\Im)}$ at each step-the cycle of the dialogue $\upsilon = 1, 2, \ldots$, in which the control object defined by the system:

$$x_i = \sum_{j=1}^{n_{\Im}} a_{jl}^{(o)} x_j + \sum_{l=1}^{m_{\Im}} b_{jl}^{(o)} \left(u_l^{(o)} + \delta u_l^{(\upsilon)} \right)$$

$$+ \sum_{q=1}^{Q_{\Im}} C_{iq}^{(o)} \left(p_q^{(\Im_o)} + \delta p_q^{(\upsilon)} \right) + \sum_{r=1}^{R_{\Im}} d_{ir}^{(o)} \theta_r^{(\Im)} \qquad (6.5)$$

it will switch from the initial state:

$$\left(x_i^{(o)}, u_l^{(o)}, p_q^{(o)}, \theta_t^{(o)}\right) \tag{6.6}$$

where index 3 is omitted, in the final:

$$\left(x_i^{(T)}, u_l^{(T)}, p_q^{(T)}, \theta_r^{(T)}\right) \tag{6.7}$$

in such a way that the totality of performance indicators

$$I_\upsilon^{(\Im)} = I_\upsilon^{(\Im)}\left(x_i^{(\Im)}\right) \to \max{}_\Sigma^{(\Im)}\left(I_\upsilon^{(\Im)}\right) \tag{6.8}$$

where $J_\Sigma^{(\Im)}\left[I_\upsilon^{(\Im)}(x)\right]$—a priori unknown utility function $DM^{(\Im)}$.

At the same time, the general scheme of the dialog mode consists of alternating phases of analysis and optimization at each step-cycle. Such steps $\upsilon = 1, 2, \ldots$ there may be several and their essence is as follows:

(1) using the information received from $DM^{(\Im)}$ at $(\upsilon - 1)$-th step the information of the computer in the automated control system forms the area of rational $u_l^{(\upsilon)}$ and $p_q^{(\upsilon)}$;
(2) calculates the state of an object $x_i^{(\upsilon)}$ and its effectiveness $I_\upsilon^{(\Im)}\left(x^{(\Im-\upsilon)}\right)$;
(3) generates some auxiliary information $И_{ЭВМ}^{(\upsilon)}$;
(4) evaluating the solution received from the computer $DM_k^{(\Im)}$ determines whether it is acceptable. If this is the case, then the dialog is finished and the DMP ends in the form of a SFA, which is transmitted to the control object. If the DM does not consider the found solution acceptable, then it remains only to analyze the auxiliary information. $И_{ЭВМ}^{(\upsilon)}$ develop your own additional information based on it $И_{DM}^{(\upsilon)}$.
(5) enters additional information into the computer $И_{DM}^{(\upsilon)}$, with which the new values are found $u_l^{(\upsilon+1)}$ and $p_q^{(\upsilon+1)}$.

As follows from the essence of the above steps, the first three of them relate to the optimization phase performed by the computer, and the last two are the analysis phase conducted by the DM. The effectiveness of such a dialog mode is determined to the greatest extent by the nature of the interaction between the DM and the computer and is quantitatively and qualitatively characterized by the additional information developed at the u step $И_{ЭВМ}^{(\upsilon)}$, $И_{DM}^{(\upsilon)}$.

A number of such dialog modes are well known and are used to develop acceptable solutions. It should only be emphasized that the common points for such dialog modes and various tasks of multi-criteria selection are:

- availability of objective and subjective data circulating through computer-DM communication channels. Moreover, the subjective part, reflecting the personal characteristics of the DM, significantly affects the entire DMP;
- the sequence of the step-by-step process of developing the $DM_k^{(\Im)}$ stable preferences based on the representation of the computer by predictive estimates of the

vector of the state of the control object and the values of indicators, while based on objective models.

When developing mathematical support for intelligent systems (IS) on the way to increase the degree of automation of the search for optimal DMP in the TSM, it is advisable to obtain an algorithm for solving the problems (6.5)–(6.8) with minimal intervention $\mathrm{DM}_k^{(\Im)}$. Let, for example, the performance indicators (6.8) are presented as a convolution:

$$I_v^{(\Im)} = \int_{\Im V}^{T}\left(\sum_{j=1}^{n_\Im}\left[\omega_{jv}^{(\Im v)}\left(x_j^{(\Im v)} - y_j^{(\Im)}\right)^2\right] + \sum_{l=1}^{m}\gamma_{lv}^{(\Im v}\left(u_l^2\right)\right)\mathrm{d}t$$

$$J_{\Sigma k}^{(\Im)} = F\left(I_{1k}^{(\Im)}, \ldots, I_{\mu k}^{(\Im)}\right) \tag{6.9}$$

where $\omega_{jv}^{(\Im v)}$ and $\gamma_{lv}^{(\Im v)}$—nonnegative coefficients for each of the indicator variants $\left(v = \overline{1, \mu^{(\Im)}}\right)$, a $y_j^{(\Im)}$; $j = \overline{(1, n_\Im)}$—the set of preset planned values of the components of the state vector of the control object.

It is clear that the appointment or choice of such coefficients indicates certain preferences $\mathrm{DM}_k^{(\Im)}$ or directly depends on his individual assessments of the importance of the output results of the state of the management object and the management costs at each v dialog mode step:

$$\begin{aligned}\omega_{jvk}^{(v)} &= \omega_{jvk}^{(v)}\left[\Psi_{kj}\left(x_j\left(t^{(v)}\right)\right)\right] \\ \gamma_{lvk}^{(v)} &= \gamma_{lvk}^{(v)}\left[\Psi\left(x_j\left(t^{(v)}\right)\right)\right]\end{aligned} \tag{6.10}$$

In the dialog mode, there is a search and modification of the DMP additives:

$$\delta u_l^{(v)}, (l = 1, \overline{m}_\Im); \delta p_q^{(v)}, \left(q = \overline{1, Q_\Im}\right); v = 1, 2, \ldots \tag{6.11}$$

to those found on $(v-1)$—step $u_l^{(v-1)}$ and $p_q^{(v-1)}$.

In such a way as to lead to the lowest values of indicators (6.9) taking into account individual estimates (6.10).

Then the general state of the TSM is characterized by the components $x_i^{(v)}, u_l^{(v)}, p_q^{(v)}, \theta_r^{(v)}$ and leading to a minimum of the functional (6.9) at the selected values of the coefficients (6.10), which can be called optimal in the sense of estimates $\mathrm{DM}_k^{(\Im)}$ (index $(\Im)$ in dependencies (6.5) and further omitted). Thus, for k-th $\mathrm{DM}_k^{(\Im)}$:

$$u_{lk}^{(v)} = u_{lk}^{(o)} + \delta u_{lk}^{(v)};\ p_{qk}^{(v)} = p_{qk}^{(o)} + \delta p_{qk}^{(v)},$$

where

$$\delta u_{lk}^{(v)} = \delta u_l(\omega(\Psi_k), \gamma(\Psi_k), x_i);$$

$$\delta p_{qk} = \delta p\big(\omega(\Psi_k), \gamma(\Psi_k), x_j\big) \tag{6.12}$$

The necessary conditions for the minimum of the functional (6.9) can be written as follows:

$$\int_{to}^{T} (X^{(v)} - Y)' \Omega_v \delta X^{(v)} dt = 0; \ \gamma_{vk} = 0 \tag{6.13}$$

or by expanding the dimension of the state vector by $m_{\mathfrak{Z}}$ component:

$$U_l = x_{n_{\mathfrak{Z}}+l} \tag{6.14}$$

where $X^{(v)} = \left(x_j^{(v)}\right), Y^{(v)} = \left(y_j^{(v)}\right)$ sign ' denotes transportation, and $\Omega_v = \left\{\omega_{ijv}\right\}_{nxn}$—symmetric, positive definite matrix formed from nonnegative coefficients ω_{ijv} так, that $\omega_{ijv} = \left\{\omega_{ijv}\right\}$ by $i = j, i \neq j$; $\delta X = (\delta x_i, \ldots, \delta x_i)'$ variations of the components of the state vector of the control object, which are from (6.5):

$$\delta X = A^{(r)} \delta X + B^{(c)} \delta U + C^{(c)} \delta P \tag{6.15}$$

$\delta X(t_0) = 0$ for arbitrary and independent variations of vector components $\delta U = (\delta u_l)'; \delta P = \left(\delta p_q\right)'$.

If, following the theory of optimal control, we introduce some transitional function $\lambda(t, t_0)$ analog of the Lagrange–Euler multiplier function defined by the control:

$$\frac{\partial \lambda(t, t_o)}{\partial t} = A\lambda(t, t_o); \ \lambda(T, t_o) = E_{nxn} \tag{6.16}$$

where matrix $A = \left\{a_{ij}\right\}$ from (6.5); E_{nxn}—a unit matrix, then a number of matrices can be calculated:

$$B = \int_{t_0}^{T} \lambda(t, \tau) B^{(0)}(\tau) d\tau; \ C = \int_{t_0}^{T} \lambda(t, \tau) C^{(0)}(\tau) d\tau;$$

where from:

$$M = (B \backslash C)', \overline{M_v} = \int_{t_0}^{T} M' \Omega_v M d dt \tag{6.17}$$

and now for the variation δX rightfully:

$$\delta X = B\delta U + C\delta P,$$

so from (6.14), taking into account independence and arbitrariness δU and δP, the necessary conditions get the form:

$$\int_{t_0}^{T} (X^{(v)} - Y^{(o)})' \Omega_v B ddt = 0; \int_{t_0}^{T} (X^{(v)} - Y^{(o)})' \Omega_v C ddt = 0 \quad (6.18)$$

where using (6.5) and (6.16), it is possible to determine $X^{(v)}$ from

$$X^{(v)} = \lambda(t, t_o)X^{(o)} + BU^{(o)} + CP^{(v)} \quad (6.19)$$

Now, given the symmetry of the matrix Ω_v and addiction (6.19) from (6.18) follows:

$$\begin{aligned} &\left(\int_{t_0}^{T} B'\Omega_v B dt\right) U^{(v)} + \left(\int_{t_0}^{T} B'\Omega_v C dt\right) P^{(v)} \\ &= \int_{t_0}^{T} B'\Omega_v \left[Y - \lambda(t, t_0)X^{(0)} - D\Theta^{(0)}\right] dt; \\ &\left(\int_{t_0}^{T} B'\Omega_v C dt\right) P^{(v)} + \left(\int_{t_0}^{T} B'\Omega_v B dt\right) P^{(v)} \\ &= \int_{t_0}^{T} C'\Omega_v \left[Y - \lambda(t, t_0)X^{(0)} - D\Theta^{(0)}\right] dt \end{aligned} \quad (6.20)$$

From where it is possible, using (6.17), to get finally:

$$\begin{pmatrix} U^{(v)} \\ P^{(v)} \end{pmatrix} = \Omega_v \left[Y - \lambda(t, t_0)X^{(0)} - D\Theta^{(0)}\right] dt \quad (6.21)$$

where $D = \int_{t_0}^{T} \lambda(t, t_0) D^{(0)}(\tau) d\tau$.

Optimal control values $U^*_{(v)}$ and parameters $P^*_{(v)}$ are now from (6.21), (6.19), (6.16) in case the matrix $\overline{M}$ it is non-singular, which practically always happens with small deviations from the program value of the components of the state vector.

The most important result of DMP optimization in the form of (6.21), (6.19), (6.16) forms the basis for further overcoming the uncertainty remaining due to the parameter $\Theta^{(o)}$ and multi-criteria $\left(v = \overline{1, \mu}\right)$.

In addition, the DM does not operate with information about the exact values of the components of the state vector $X = \{x_i\}'$, $\left(i = \overline{1, n_{\Im}}\right)$, a receives information about the measured and calculated values of the components (Fig. 3.8), for example, in the form:

$$Z_j^{(v)} = \sum_{i=1}^{n_3} g_{ji} x_i^{(v)} + \alpha_j^{(v)}, v = 0, 1, 2, \ldots \text{ – steps} \quad (6.22)$$

where $x_i^{(v)}$—random component of the state vector:

$$X^{(\upsilon)} = AX^{(\upsilon)} + BU^{(\upsilon)} + CP^{(\upsilon)} + D\Theta^{(\upsilon)} + \zeta^{(\upsilon)}, X^{(\upsilon)}(t_o) = X^{(0)}(t_0 \le t \le T) \tag{6.23}$$

g_{ij}—a given matrix of coefficients;

$\alpha_j^{(\upsilon)}, \zeta_j^{(\upsilon)}$—components of uncorrelated random processes modeling the results of disturbing effects on the control object (6.23) and on the measurement process (6.22). The zero value of their mathematical expectation and the set values of the autocorrelation matrices are assumed:

$$K_\zeta(t, \tau), K_\alpha(t, \tau) \tag{6.24}$$

Initial state of the object $X^{(\upsilon)}(t_o)$ set by the value of the mathematical expectation $X^{(o)} = \left\{x_i^{(o)}\right\}'$ and the correlation matrix $K^{(o)}$, not correlated with other random processes.

Now accumulating measurable information $Z^{(o)}$, $Z^{(1)}$,..., $Z^{(\upsilon)}$,..., as well as receiving status assessments $\hat{X}^{(o)}$, $\hat{X}^{(1)}$,..., $\hat{X}^{(\upsilon)}$,... according to (6.22)–(6.24) and evaluation of its effectiveness:

$$I_{\text{v}}^{(\upsilon)}(\hat{X}, \hat{U}, \hat{P})^{(\upsilon)}, (\upsilon = 0, 1, \ldots), (\upsilon = \overline{1, \mu}),$$

where

$$I_{\upsilon}^{(\upsilon)} = I^{(\upsilon)}\left\{\hat{X}[\Omega_\upsilon(\Psi_K)], \hat{U}[\Omega_\upsilon(\Psi_K)]\right\} \tag{6.25}$$

An interactive complex, including an automated control system, with elements of an IS, working in an interactive mode directly with the DM or through a consultant or their group, determines the sequence

$$\left[\delta U^{(\upsilon)}, \delta P^{(\upsilon)}\right], (\upsilon = 0, 1, \ldots)$$

correction of optimal values $\left[U^{(\upsilon)}, P^{(\upsilon)}\right]$, found from (6.21), which will ensure the convergence of the process $\left[\hat{X}^{(\upsilon)}, \hat{U}^{(\upsilon)}, \hat{P}^{(\upsilon)}\Theta^{(\upsilon)}\right]$ to a set acceptable for a given k-th $\text{DM}_k[X(T) \approx Y, U(T), P(T), \Theta(T)]_{\upsilon k}$ according to the totality of indicators $I_{\upsilon k}, (1, \mu)$.

Optimal filtering algorithms are successfully used to obtain an estimate of the state vector and its forecast.

Acceptable for DM_k the solution is a choice of such values $\omega_{\upsilon k}(\psi_k)$, which on the surface of pareto-optimal solutions will indicate an area or point corresponding to individual estimates DM_k (Fig. 6.12).

One of the possible ways, starting from a volitional act of choice and up to conducting an expert survey, can be found in [1–3]. It is important to emphasize only that, due to the presence of mathematical models for the process of changing the state vector of the control object and measuring devices in the structure of the TSM, it is

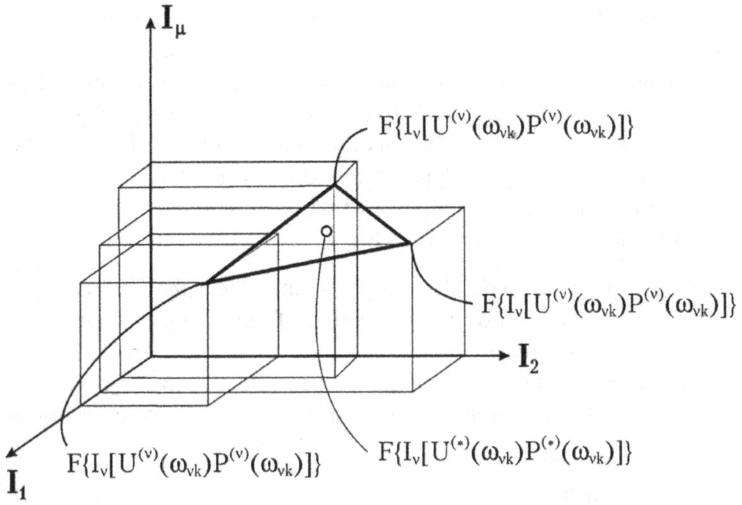

Fig. 6.12 Pareto-optimal surface or area of optimal values of performance indicators depending on the values Ω_k

possible to build an optimization algorithm implemented in the automated control system with IS, which leads to a significant narrowing of the surface $\Pi(U^*, P^*)$. Therefore, the differences $F^{(\upsilon)}(\upsilon = 0, 1, \ldots)$ Often, in the case considered here, the development of a DMP at TPM is not of a fundamental nature. Moreover, the independence of the class of the function of changing the vector of the state of the control object under optimal control (6.21) from the method of convolution (6.9) of the vector efficiency indicator is known.

And in problematic cases of the course of DMP with TPM, a pair of private indications of the “accuracy-time” type is most often used [3, 4], which almost always occurs at the lower level of TPM.

The analytical hierarchy method occupies a special place when supporting the decision-making process in unstructured transport processes.

In the case of a small number of specified alternatives, it seems reasonable to direct the efforts of the DM to compare only the specified alternatives. It is this idea that underlies the analytical hierarchy method (AHM), the author of which is T. Saati [1].

AHM is more often used when solving group II tasks, i.e. creative, non-programmable tasks.

The statement of the problem solved using this method is usually as follows.

It is known: the general goal (goals) of solving the problem; N criteria for evaluating alternatives; n alternatives.

Required: choose the best alternative.

AHM consists of several stages:

1. Structuring the task in the form of a hierarchical structure with several levels: goals—criteria—alternatives.
2. Implementation of DM paired comparisons of elements of each level, the results of the comparisons are translated into numbers using a special table.
3. Calculation of importance coefficients for elements of each level. At the same time, the consistency of the judgments of the DM is checked.
4. Calculation of a quantitative indicator of the quality of each of the alternatives and determination of the best alternative.

Let's consider such a problem. The transport committee of the City Administration N is faced with the task of choosing the option of purchasing vehicles to provide urban passenger transportation.

Four possible procurement options have been selected: $З_1$, $З_2$, $З_3$ и $З_4$. The evaluation of each of the options is supposed to be carried out according to three criteria:

- total cost of vehicles—*C*;
- maintenance and repair costs—*P*;
- the amount of emissions of harmful substances—*B*.

It is clear that the assessment for each of the criteria should be of as little importance as possible. But among the proposed options, there is no one whose estimates according to individual criteria would simultaneously take the smallest values. It is necessary to choose the best option for purchasing vehicles.

I. The structure of the problem being solved.

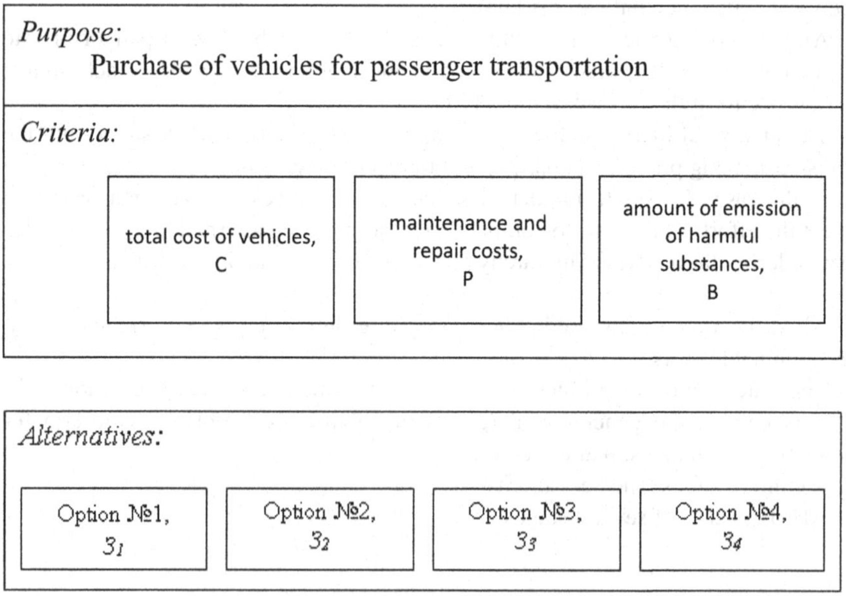

Table 6.3 Scale of verbal definitions of the level of importance

Verbal description of the importance level	Quantitative value
Equal importance	1
Moderate superiority	3
Substantial or strong superiority	5
Significant (large) superiority	7
Very great superiority	9

II. For the purpose of pairwise comparisons, a scale of verbal definitions of the level of importance is given to the DM, each of which is assigned a number (Table 6.3):

When comparing elements belonging to the same level of the hierarchy, the DM expresses its opinion using one of the definitions given in Table 6.3. If desired, the DM can use even integers, expressing intermediate levels of preference in importance. The corresponding numbers are entered into the comparison matrix.

Let the matrix of comparisons of criteria for choosing a TSy procurement option look like this (Table 6.4).

At the lower level of the hierarchical scheme, the specified alternatives (procurement options) are compared—$З_j$) for each criterion separately.

Let the corresponding comparisons be given in Tables 6.5, 6.6 and 6.7:

III. Tables 6.5, 6.6 and 6.7 allow you to calculate the coefficients of importance of the corresponding elements of the hierarchical level. To do this, you need to

Table 6.4 Matrix of comparisons of criteria for choosing a procurement option

Criteria	C	$З$	B	Eigenvector $a_i = \left[\prod_{j=1}^{3} a_{ij}\right]^{1/3}$	Criterion weight $\omega_i = \frac{a_i}{\sum_{j=1}^{3} a_j}$
C	1	5	4	2.714	0.680
$З$	1/5	1	3	0.843	0.211
B	1/4	1/3	1	0.437	0.109

Table 6.5 Comparison by criterion C

Alternative	$З_1$	$З_2$	$З_3$	$З_4$	Eigenvector $b_i = \left[\prod_{k=1}^{3} b_{ik}\right]^{1/4}$	Criterion weight $\omega_i = \frac{b_i}{\sum_{k=1}^{4} b_k}$
$З_1$	1	1/5	1/7	1/9	0.237	0.0393
$З_2$	5	1	1/3	1/5	0.760	0.126
$З_3$	7	3	1	1/3	1.63	0.270
$З_4$	9	5	3	1	3.41	0.565

Table 6.6 Comparison by criterion *З*

Alternative	$З_1$	$З_2$	$З_3$	$З_4$	Eigenvector	Criterion weight
$З_1$	1	0.111	0.200	0.143	0.237	0.0450
$З_2$	9	1	3	1	2.280	0.432
$З_3$	5	0.333	1	1	1.14	0.215
$З_4$	7	1	1	1	1.63	0.308

Table 6.7 Comparison by criterion *B*

Alternative	$З_1$	$З_2$	$З_3$	$З_4$	Eigenvector	Criterion weight
$З_1$	1	3	5	9	3.409	0.565
$З_2$	0.333	1	3	7	1.627	0.270
$З_3$	0.200	0.333	1	5	0.76	0.126
$З_4$	0.111	0.143	0.200	1	0.24	0.0393

calculate the eigenvectors of the matrix, and then normalize them. The components of the eigenvector are defined as the geometric mean of the matrix elements in the corresponding row.

The values of the eigenvector and importance coefficients are given in the last two columns of each table.

When filling in the matrices of pairwise comparisons, a person can make mistakes. For example, a violation of transitivity (from the fact that $a_{ij} > a_{jk}, a_{jk} > a_{js}$ you shouldn't $a_{ij} > a_{js}$), violation of the consistency of numerical judgments $\left(a_{ij} \cdot a_{jk} \neq a_{ik}\right)$.

To detect inconsistency, you can use the procedure for calculating the consistency index of judgments, carried out according to the matrix of paired comparisons and consisting in the following:

1. In the matrix of paired comparisons, the elements of each column are summed up.
2. The sum of the elements of each column is multiplied by the corresponding normalized components of the weight vector determined from the same matrix.
3. The obtained numbers are summed up, the sum is denoted as $\lambda_{\max}$.
4. The consistency index is found: $L = \frac{\lambda_{\max}-n}{n-1}$, where n—the number of elements being compared, i.e. the size of the matrix (for a skew-symmetric matrix $\lambda \geq n$).
5. The average value of the consistency index R is calculated for matrices filled in randomly. So, for a matrix of size $n = 7$, the index $R = 1.32$, and for a matrix of size $n = 8$, the index R = 1.41.

The consistency ratio is calculated: $T = \frac{L}{R}$. When applying the method, the level is considered desirable $T \leq 0, 1$. If the value of T exceeds this level, it is recommended to make comparisons again.

IV. The synthesis of the obtained importance coefficients is carried out according to the formula:

$$S_j = \sum_{i=1}^{N} \omega_i \cdot V_{ji},$$

where S_j—quality indicator of the j-th alternative; ω_i—weight of the i-th criterion; V_{ji}—the importance of the j-th alternative according to the i-th criterion.

For the procurement options under consideration:

$$S_1 = 0.0393 \cdot 0.680 + 0.0450 \cdot 0.211 + 0.565 \cdot 0.109 = 0.0980;$$

$$S_2 = 0.126 \cdot 0.680 + 0.432 \cdot 0.211 + 0.270 \cdot 0.109 = 0.206;$$

$$S_3 = 0.270 \cdot 0.680 + 0.215 \cdot 0.211 + 0.126 \cdot 0.109 = 0.242;$$

$$S_4 = 0.565 \cdot 0.680 + 0.308 \cdot 0.211 + 0.0393 \cdot 0.109 = 0.453.$$

So, procurement option No. 4 is recognized as the best.

Also known as the proposed by Professor F. Lutsmoy is a multiplicative method of analytical hierarchy, which has a slightly different methodological justification.

It is based on two main provisions.

Firstly, if the DM determines the relations (and not the absolute values) of two elements of the corresponding hierarchy level, then it is more logical to multiply such relations than to sum up the values obtained from comparisons.

Secondly, the transition from verbal comparisons to numbers should take place on the basis of some assumptions about human behavior in comparative measurements.

6.3 Modeling of the Evaluation of the Effectiveness of DMP in TPM

At various levels of the hierarchical structure of the TPM, specific conditions of uncertainty arise, and input information based on which $DM^{(\Im)}$ makes decisions, as a rule, contains interference and distortion. In these conditions typical for practical activity $DM^{(\Im)}$ a special role is played by the noise immunity and prognostic feature of the complex, which is based on an automated control system with an IS with a database of incidents and their circumstances and a knowledge base for their prevention and DM with a group of research consultants who participate in the development of support for DMP and at the training stage of an automated control system or IS [5–7].

The most important issue in assessing the activities of such a complex is the modeling of the most likely outcomes that are the result of the DMP developed by it. This would make it possible to trace the most probable transitions in the space of the vector of the state of the control object, for example, in the form of an assessment of the effectiveness of the DMP in the TPM based on the prediction of the most likely such transitions. $p_{ij}\left(u_{ij}^{(\mathfrak{Z})}\right)$ from (Fig. 3.4).

Such models make it possible to evaluate the effectiveness of not only the decisions already made, but also to evaluate their alternatives.

Let be a set of parameters of a dynamic transport situation and commands $\mathrm{DM}^{(\mathfrak{Z})}$ for this particular section of the movement, the state vector is determined $X_i^{(\mathfrak{Z})}$, reflecting the situation $S_r^{(\mathfrak{Z})}$, where $i = \overline{1, n_{\mathfrak{Z}}}$ and $r = \{1, 2, \ldots, K_{\mathfrak{Z}}\}$—a finite number of such situations, for each of which is known, for example, the distribution function of the time spent in the situation d before the transition to the state-situation j:

$$F_{r_j};\ F_{r_j}(0) = 0$$

In addition, the elements of the transition probability matrix are given:

$$P = \left\| p_{r_j} \right\|$$

The whole set of states X_i can be broken down into a variety of acceptable situations $G^{(\mathfrak{Z})} : \left\{S_r^{(\mathfrak{Z})}; r = \nu + 1, \ldots, K_{\mathfrak{Z}}\right\}$ and unacceptable, in which there are traffic accidents (TA) or prerequisites for them (PTAP) $H^{(\mathfrak{Z})} : \left\{S_r^{(\mathfrak{Z})}; r = \nu + 1, \ldots, K_{\mathfrak{Z}}\right\}$. At the same time, it is assumed that the types of unacceptable (or failed) situations are known and the number of them—$\overline{\mu_{\mathfrak{Z}}}$:

$$K_3 = v + (\mu_3 - 1)$$

Based on the analysis of statistical characteristics, the total probabilities of the state for the "exit" from the set from any state can be found x_i:

$$\varepsilon_i = \sum_{j \in H} p_{ij}, i \in G \tag{6.26}$$

If we now also evaluate statistical characteristics such as the particulars of transitions within the set and the average values of the time before such transitions $\hat{\tau}_{ij}^{(0)}$; $(i, j \in G)$ then you can find the values:

(1) average weighted failure rate:

$$\lambda = \sum_{i=1}^{m} \frac{\varepsilon_i}{\hat{\tau}_{ij}^{(o)}} \tag{6.27}$$

(2) the average time of the "return" of the process to the state $x_i \in G$:

$$\hat{\tau}_{ij}^{(o)} = \frac{1}{p_i^{(0)}} \sum_{l \in G} \overline{p}_l^{(0)} \tau_l^{(0)} \tag{6.28}$$

(3) the unconditional residence time of process X in the state «l»:

$$\tau_l^{(0)} = \sum_{j \in G} p_{lj}^{(o)} \tau_{kj}$$

In the expression (6.28) $\overline{p}_i^{(o)}$—stationary state probability for a Markov chain:

$$P^{(o)} = \left\| p_{ij}^{(o)} \right\|, (i, j \in G)$$

Now we can estimate one of the most interesting characteristics of the causal chain—the probability of the process leaving the core of the set G at the time t

$$P(\tau \leq t) \cong 1 - \ell^{-\lambda t} \tag{6.29}$$

Number $c \in \mu \in H$ failure states and the probability of their occurrence is a quantitative measure of the load for subsequent levels of the hierarchy in the TPM. Thus, the effectiveness of decision-making processes at any of the levels $\Im$ can be estimated by the number of failures $c^{(\Im)} \in \mu^{(\Im)} \in H^{(\Im)}$ the magnitude of the probabilities of occurrence for which is "tangible", i.e.

$$P\left(\tau^{(\Im)} = T^{(\Im)}\right) \geq k^{(\Im)}(0) \tag{6.30}$$

where $T^{(\Im)}$—the specified calendar period for this $\Im$ level;

$k^{(\Im)}(0)$—the acceptable warranty value of the probability of failure.

The presence of the listed statistical data allows us to assume that there is a possibility in the development of an adaptive complex for predicting transitions in the state space on $\Im$ at the TPM level. Along with the overall probability of exit (6.30), which assesses the effectiveness of any action DM$^{(\Im)}$ relevant $\Im$ levels you can enter an estimate of the probability of such an output for each specific type of TA or PTA:

$$k^{(\Im)}(0), c = \overline{1, m}\text{, где } i \in G; j \in H.$$

In this case, the computer memory (IS) of the complex captures the final vertices of the graph connecting the situations-the vertices of the set G with the vertices of the set N. Now the request can be answered in the form of a Boolean vector:

$$\delta^{(3)} = \left(\delta_1^{(3)}, \ldots, \delta_m^{(3)}\right); \delta_j^{(3)} = \begin{cases} 1 - \text{the most likely } j \text{ - th event for the level 3;} \\ \qquad\qquad 0 - \text{otherwise;} \end{cases} \tag{6.31}$$

and forecast:

$$\pi^{(3)} = \left(\pi_1^{(3)}, \ldots, \pi_m^{(3)}\right); \pi_j^{(3)} = \begin{cases} 1 - \text{the most likely } j \text{ - th event for the level 3;} \\ \qquad\qquad 0 - \text{otherwise;} \end{cases}$$

Value 1 or 0 for vector components $\delta^{(\mathfrak{Z})}$ it can be obtained, for example, using (6.31). Then the set H is defined for each level -$H^{(\mathfrak{Z})}$.

If we compare a couple of values to each situation $\left(\chi_s^{(\mathfrak{Z})}, \varphi_s^{(\mathfrak{Z})}\right)$ real nonnegative numbers, where $\chi_s^{(\mathfrak{Z})}$—the coefficient of "severity" of the condition (X_s) for this particular s-th traffic accident TP (PTP), $\varphi_s^{(\mathfrak{Z})}$—the "training" coefficient defined below.

Let the "severity" of TA (PTA) be ranked in accordance with the initial (a priori) value of the coefficients $\chi_s^{(\mathfrak{Z})}$ and $\varphi_s^{(\mathfrak{Z})}$.

The operation of the complex can be determined in two modes: the working mode, when they receive answers to requests for evaluation of the effectiveness of decisions made and the forecast; and in the training mode, when the working mode is repeated with subsequent correction of a pair of values $\chi_s^{(\mathfrak{Z})}$ and $\varphi_s^{(\mathfrak{Z})}$ for everyone $\overline{\mu}_{\mathfrak{Z}}$ one into the other, and form at least one path connecting ρ and η and containing all situations j:

$$\delta_j^{(\mathfrak{Z})} = 1$$

For the researcher, the evaluation of three types of path is of the greatest interest, when the transition leads to a return to the initial state, when to the set G, and when to the most severe TA (PTA) of the set $H^{(\mathfrak{Z})}$. Let's introduce the function:

$$\Psi_c^{(3)} = \sum_{j \in l_k^{(\delta)}(p,n)} \omega_{jc}^{(3)} z_j - \sum_{j \notin l_k^{(\delta)}(p,n)} \omega_{jc}^{(3)} z_j \tag{6.32}$$

where

$$c \in H^{(3)}; \omega_{jc}^{(3)} = \begin{cases} 1 - \text{with probability } p_k \text{ the fact that } j \text{ and } c \\ \quad - \text{ the beginning and the end of the path} \\ 0 - \text{ with probability } 1 - p_k \end{cases}$$

In addition, we will set similar values for each situation and the level of "training" coefficients $\varphi_s^{(\mathfrak{Z})}$. Then there are three possible transitions from each i-th situation—to situations j and c or a return to the i-th. In accordance with this, each of the paths can be associated with a random variable $\xi_{ir}^{(\mathfrak{Z})}$, where $(r = j, c, i)$, depending on the magnitude of the difference:

$$\xi_{ic}^{(3)} = \xi_{ic}\left(\Psi_c^{(3)} - \varphi_c^{(3)}\right) \tag{6.33}$$

and increment values $\Delta\xi$, defined by the sign:

$$\Delta\xi_i = \xi_{ij}^{(3)} - \xi_{ic}^{(3)} \begin{cases} \geq 0 - \pi_j^{(3)} = 1 - forecast\ of\ transition\ to\ the\ state\ j \\ = 0 - \pi_i^{(3)} = 1 - forecast\ of\ transition\ to\ the\ state\ i \\ < 0 - \pi_c^{(3)} = 1 - forecast\ of\ transition\ to\ the\ state\ c \end{cases}$$

Training can now be conducted in such a way that it occurs from both directions, i.e. that the values of the quantities change z_j and (6.32) and values $\varphi_c^{(\Im)}$. For z_j and $\varphi_j^{(\Im)}$ here then we can propose the following iterative procedure on υ iterations:

$$Z_j(v,\delta) = \begin{cases} Z_j(v) + 1, if\ \ j \in l_3^{(\delta)}(p,n), Z_j(v) \neq K - the\ specified\ number \\ Z_j(v) - 1, если\ j - the\ beginning\ of\ the\ path, the\ end\ of\ which\ belongs\ to\ l_3^{(\delta)}(p,n), Z_j(v) \neq 0 \\ Z_j(v) - in\ all\ other\ cases \end{cases} \tag{6.34}$$

Similarly for $\varphi_j^{(\Im)}(\upsilon)$.

Considering the value of the corresponding probabilities of using training procedures (6.34) for $\varphi_j^{(\Im)}(\upsilon, \delta)$. and $Z_j(\upsilon, \delta)$. for the number of iteration steps $\upsilon = R$ it is possible to find the value of the probabilities of the correctness of the forecast for each of the levels:

$$P^{(3)}\{\delta \in \pi\} = \Pi\Phi_{i,\delta_i=1}\left\{\frac{\hat{\xi}_{ci}^{(3)} - \hat{\xi}_{cj}^{(3)}}{\left[\sigma\left(\xi_{ci}^{(3)}\right) -\right]}\right\} P\left[c \in l_s^{(\delta)}(p,n)\right]$$
$$+ \Pi\Phi_{i,\delta_i=1}\left\{\frac{P_{ci} * P_c^{(\delta)}(R) - P_{sj} * P_j^{(\delta)}(R)}{\left[\left(P_{ci} * P_i^{(\delta)} + P_{cj} * P_j^{(\delta)}\right)\left(1 - P_{ci} * P_i^{(\delta)} + P_{cj} * P_j^{(\delta)}\right)\right]^{\frac{1}{2}}}\right\}$$
$$* P\left[c \in l_s^{(\delta)}(p,n)\right] \tag{6.35}$$

where $\Phi\{\cdot\}$—density of the normal distribution law;

$\hat{\xi}$—the average value of a random variable;

$\sigma(\xi)$—its variance;

$P_i^{(\delta)}$—the probability that the i-th situation belongs to a learning set whose step length does not exceed R;

By determining the values (6.35) for a given number R, it is possible, given a certain guarantee value $R^{(\Im)}(R)$ determine the degree of completion of the learning process or, conversely, the need for its resumption.

Having carried out periodic training of the complex by changing the corresponding parameters in it in accordance with the values of the probability of correctness of the forecast (6.35), it is thus possible to carry out not only a posteriori assessment of the effectiveness of decision-making processes (6.30), but also to predict state transitions and their probabilities. Naturally, at the same time, it is also necessary to simulate the process of generating the states of the control object according to the decisions made, which was postulated here.

6.4 Structural Analysis of the TPM and the Principle of the Study of DMP Taking into Account the Human Factor

All phenomena, objects, processes existing in this world can, roughly speaking, be divided into two groups, the first of which unites everything created by nature (God, if you like), and the second—everything that is created by man. All vehicles and paths are the roads of their route, as well as the organization and management of their movement, i.e. all types of TPM structures also belong to the second group.

A characteristic feature of the second group is mandatory expediency, i.e. the availability of performance indicators. So, in relation to the transport system, it is a given movement of goods and passengers at the lowest cost of energy, space, time or any newly discovered other resources. If we talk about such an important part of transport as a traffic management system, such indicators include the efficiency of using space (air, water, land) for organizing, planning and managing safe, regular and economical traffic. This is most often explicated in the form of optimization tasks, such as maximizing throughput $\mu(i, t)$ space elements with specified restrictions on violations of regularity conditions, on allowable costs, strict security conditions, etc. $L_{i\alpha}^{(\Im)}(t)$:

$$\max\left\{\mu_{\alpha}^{(\Im)}\left(i^{\Im}, t\right)/L_{i\alpha}^{(\Im)}(t)\left(i^{\Im}, t\right) \geq L_{i\Im\alpha}^{*}\right\}$$

where as an argument on each $\Im$ at the TPM level, the main product produced by the management system is information. Thus, for a management system, such a characteristic feature results in the presence of an optimization task of maximizing the information produced at a given level of its quality—error-free, timely, presentation form, etc. All of the above suggests the existence of one of the most important principles of the functioning of the transport management system—the principle of its economic feasibility: For example, the necessary increase in throughput is caused by the inevitable need to increase the amount of information in the organization, planning and direct management.

At the same time, the economic costs of additional operations for organization, planning and management should be compensated by revenues from increased transportation, reducing their danger, and improving their quality. That is, the value of information produced by the management system and determined by all costs is, on the other hand, strictly dependent on the income to which it leads availability for transportation with the achievement of a given capacity.

The presence of such a principle in the functioning of the system requires, in turn, its use in the study of DMP in the TPM. Therefore, it is advisable to show at least a fragmentary demonstration of the productivity of its application by the example of the study of some processes in such a type of control system, for example, as a space use and motion control system. At the same time, applying the principle of such a general nature, although one cannot hope for the specifics of the results, one can be sure of maintaining their generality for control systems in any types of transport or in intermodal transport networks and systems having a complex character (intertransport hubs, e.g., or systems for traffic management in emergency situations).

Indeed, any of the systems of this kind can be attributed to large or complex organizational systems, where a significant role is played by the established traditions, assessment, style of activity of collectives. The rules, methods and methods developed at the same time are often (almost always) the result of trial and error and, ultimately, in most cases, provide rational processes and DMP, including in the system. Hence, an approach of process analysis based on solving the inverse problem of process optimization can be considered common to all of them, when it is obviously considered optimal and the task is to find (formalize) an efficiency indicator, in the sense of which optimality and improve the process under study. Similarly, the situation is taking into account the ergoticity of these systems, i.e. taking into account those properties that are provided by the presence of human collectives and human operators in them in any modern and projected transport management system for the foreseeable future.

Having defined the structure of the TPM as an organizational one, the purposeful nature of its functioning is assumed from the very beginning. The formalization of goals, taking into account the assessments of the behavior of DM teams, is associated with a number of factors of a prestigious, moral, social, economic, political and other nature. Comprehensive accounting of all such factors in mathematical models is hardly possible yet. One of the most significant components of this list of factors is the economic (material, economic) interests of specialists—DM at the DMP in the TPM.

The identification and consideration of these interests would make it possible to identify and explore at least fragments of one of the most important components of the functioning of the transport system—its economic management mechanism. At the same time, it is necessary to formalize the planned indicators and indicators for assessing the real state of the system elements, bringing them into line with the existing provisions on methods of economic stimulation.

It is clear that, given the pioneering nature of the attempt, she can be forgiven for some simplified approach. However, even within the framework of such a simplified approach, when a number of conditions and characteristics of the system remain not considered, it is safe to say that the principle of taking into account human and personal factors is included in the study of DMP in TPM.

The bottom line here is that the main product of any management system—information—is produced mainly by a person-a specialist in direct management on each of the $\Im$ TPM levels. Each such specialist or a team of such specialists $\mathrm{DM}_k^{(\Im)}\left(k=\overline{1,N_\Im}\right)\left(k=\overline{1,N_\Im}\right)$, where.

$N_\Im$—the number of such specialists on $\Im$ at the same level, they form an active element whose interests do not always coincide with the interests of the same collectives working in parallel, and with the interests of higher level $(\Im+1)$ at the level of the authorities-the center or $\mathrm{DM}^{(\Im+1)}$. The principle of taking into account human and personal factors in the study of DMP at TPM has already been discussed above; here we consider an aspect that consists precisely in building models of rational economic functioning of the system, which interests all elements-teams of specialists in the results of its activities $DM_k^{(\Im)}\left(k=\overline{1,N_\Im}\right)$.

Any variant of the control system can be represented as a combination of a set of two-level fan type: one center and a number of similar executive elements. A typical example of the structure of such a two-level system is the one shown in (Fig. 3.5).

Let be the number of executive elements on the lower $\Im$ level—$N_\Im$ and each of them is assigned its own task or its own work plan (in units, e.g., serviced vehicles or other type of activity) $f_{jk}^{(\Im)}(y,t)$, $j=\overline{1,n_y})$—type of vehicles or type of work; $k=\overline{1,N_\Im}$ and costs costs$\gamma_k^{(\Im)}(t)$ to fulfill the plan $\sum_{j=1}^{m} f_{jk}^{(\Im)}(y,t)$; moreover, their minimum value can be defined as /4/:

$$\gamma_k^{(\Im)} \geq \min \gamma_k^{(\Im)} = \frac{1}{2r_k^{(\Im)}} \sum\nolimits_{j=1}^{m} \left[f_{jk}^{(\Im)}(y,t)\right]^2 = \gamma_k^* \tag{6.36}$$

where $r_k^{(\Im)}$—efficiency coefficient of the k-th element $\mathrm{DM}_k^{(\Im)}$ on $\Im$ level.

It is clear that the plan and costs by type of work and by the number of performers can be estimated as:

$$\min\left\{\gamma(\Im)/\prod = \sum_{k=1}^{N_\Im} \prod_{k}^{(1)} \approx \sum_{k=1}^{N_\Im} \sum_{j=1}^{n_{y\Im}} f_{jk}(y,t)\right\} \tag{6.37}$$

where Π—a general work plan, the implementation of which is important while minimizing total costs:

$$\min\left\{\gamma^{(\Im)}/\Pi = \sum_{k=1}^{N_\Im} \Pi_k^{(1)} \approx \sum_{k=1}^{N_\Im} \sum_{j=1}^{n_{y\Im}} f_{jk}(y,t)\right\} \tag{6.38}$$

The interests of each of $\mathrm{DM}_k^{(\Im)}$ not always and not in everything coincide with the interests $\mathrm{DM}_k^{(\Im+1)}$ and they are dictated by material, prestigious, psychological and other motives. If we define the price in terms of the final factor $\mathrm{DM}_k^{(\Im-1)}$—costs C_{ik} for each j-th type of work, then the interest is understandable (explicable) $\mathrm{DM}_k^{(\Im)}$ in achieving maximum income $D_k^{(\mathfrak{Z})}$ (subject to the fulfillment of their plan (no fines):

$$\max\left\{\left(\sum_{j=1}^{n_{y\mathfrak{Z}}} C_{jk} f_{jk}(y,t) - \gamma_k^{(\mathfrak{Z})}\right)/P_i = \sum_{j=1}^{m} x_{ji}\right\}$$
$$= \max\left\{D_k^{(1)}/\Pi_k^{(1)}\right\} \tag{6.39}$$

However, in optimization problems $\mathrm{DM}_k^{(\mathfrak{Z}+1)}$ and elements $\mathrm{DM}_k^{(\mathfrak{Z})}$ cost estimates are not always known C_{ik}, efficiency coefficient of activity k-th element $r_k^{(\mathfrak{Z})}$ (sometimes with more detail, estimates are required $r_{jk}^{(\mathfrak{Z})}$—for each type of activity) and planned values can also be assigned $\mathrm{DM}_k^{(\mathfrak{Z}+1)}$ and be consistent with $\mathrm{DM}_k^{(\mathfrak{Z})}$. Usually, the terms used for this are the characteristics of the bandwidth of each $\mathrm{DM}_k^{(\mathfrak{Z})}\left(k = \overline{1, N_{\mathfrak{Z}}}\right)$ and intensity: $\mu_k^{(\mathfrak{Z})}$, $\overline{N}_{kj}^{(\partial on)}$ и $\lambda_j^{(\mathfrak{Z})}$, $\overline{N}_j^{(\mathfrak{Z})^*}$ (sometimes—$\mu_{jk}^{(\mathfrak{Z})}$ and $\lambda_{jk}^{(\mathfrak{Z})}$). In this case, in tasks (6.38), (6.39) it is necessary to take into account the constraints of the form:

$$J_{\Sigma}^{(1)} = \sum_{k=1}^{N_{\mathfrak{Z}}} \frac{1}{2\hat{r}_k} \sum_{j=1}^{m} [f_{jk}^{(\mathfrak{Z})}(y,t,\hat{r}_k)]^2 - min_f,\ \text{при} \sum_{j=1}^{n_{y\mathfrak{Z}}} \sum_{k=1}^{N_{\mathfrak{Z}}} f_{jk}^{(\mathfrak{Z})}(y,t,\hat{r}_k) = \Pi^{\mathfrak{Z}}$$
$$r_k^{(\mathfrak{Z}m)} \le \hat{r}_k^{(\mathfrak{Z})} r_k^{(\mathfrak{Z}\mu)};\ k = \overline{1, N_{\mathfrak{Z}}}, j = \overline{1, n_{y\mathfrak{Z}}} \tag{6.40}$$

where k—number of hours per work shift $\mathrm{DM}_k^{(\mathfrak{Z})}$;

and where $\overline{N}_{kj}^{(\partial on)}$—permissible number of load elements for the k-th $\mathrm{DM}_k^{(\mathfrak{Z})}$;

$\overline{N}_j^{(\mathfrak{Z})^*}$—the possible number of elements of such a load.

Let what is known $\mathrm{DM}_k^{(\mathfrak{Z}+1)}$—coefficient estimation $\hat{r}_k^{(\mathfrak{Z})}$ and $\hat{R}^{(\mathfrak{Z})} = \left\{\hat{r}_k^{(\mathfrak{Z})}\right\}$ and on their basis, a rational definition of the plan is required—$f_{ik}(y,t)\left(\hat{r}_i\right)$ and prices—$C_i\left(\hat{r}_i\right)$. Thus, if the control function is determined by the condition of rigid centralization in control systems, in which $\mathrm{DM}_k^{(\mathfrak{Z}+1)}$ optimizes the overall performance for the entire system, then the following task plans are appropriate—$f_k^{(\mathfrak{Z})}\left(\hat{r}_i\right)$, at which the total costs will be the least:

$$L^{(\mathfrak{Z})}(f,\ell) = \sum_{k=1}^{N_{\mathfrak{Z}}}\left(\frac{1}{2\hat{r}_k^{(\mathfrak{Z})}}\sum_{j=1}^{m}\left[f_{jk}^{(\mathfrak{Z})}(y,t,\hat{r}_k^{(\mathfrak{Z})})\right]\right)^2 - \ell\left(\sum_{j=1}^{n_{y\mathfrak{Z}}}\sum_{k=1}^{N_{\mathfrak{Z}}} f_{jk} - \Pi\right) \therefore$$
$$\frac{\partial L}{\partial f_{jk}} = 0 \therefore \sum_{j=1}^{m} f_{jk}^{*}(\hat{r}) = \ell\hat{r}_k \therefore \sum_{k=1}^{N_{\mathfrak{Z}}}\sum_{j=1}^{n_{y\mathfrak{Z}}} f_{jk}^{*}(\hat{r}_k) = \ell\sum_{k=1}^{N_{\mathfrak{Z}}} \hat{r}_k^{(\mathfrak{Z})} = \ell\hat{R}^{(\mathfrak{Z})} \therefore$$
$$l = \frac{\Pi}{\hat{R}} \therefore \sum_{j=1}^{n_{y\mathfrak{Z}}} f_{jk}^{*(\mathfrak{Z})}\left(y,t,\hat{r}_k^{(\mathfrak{Z})}\right) = \hat{r}_k^{(\mathfrak{Z})}\frac{\Pi^{(\mathfrak{Z})}}{\hat{R}^{(\mathfrak{Z})}} \tag{6.41}$$

The solution of the problem (6.41) is relatively easy to find by the Lagrange method:

$$L^{(\Im)}(f,\ell)=\sum_{k=1}^{N_\Im}\left(\frac{1}{2\hat{r}_k^{(\Im)}}\sum_{j=1}^{m}\left[f_{jk}^{(\Im)}\left(y,t,\hat{r}_k^{(\Im)}\right)\right]\right)^2-\ell\left(\sum_{j=1}^{n_{y\Im}}\sum_{k=1}^{N_\Im}f_{jk}-\Pi\right)\therefore$$

$$\frac{\partial L}{\partial f_{jk}}=0\therefore\sum_{j=1}^{m}f_{jk}^{*}(\hat{r})=\ell\hat{r}_k\therefore\sum_{k=1}^{N_3}\sum_{j=1}^{n_{y\Im}}f_{jk}^{*}(\hat{r}_k)=\ell\sum_{k=1}^{N_\Im}\hat{r}_k^{(\Im)}=\ell\hat{R}^{((\Im)}\therefore$$

$$\ell=\frac{\Pi}{\hat{R}}\therefore\sum_{j=1}^{n_{y\Im}}f_{jk}^{*(\Im)}\left(y,t,\hat{r}_k^{(\Im)}\right)=\hat{r}_k^{(\Im)}\frac{\Pi^{(\Im)}}{\hat{R}^{(\Im)}} \tag{6.42}$$

with: $\sum_{j=1}^{n_{y\Im}}f^{*(\Im)}\left(y,t,\hat{r}_i\right)\le\mu_{jk}^{(\Im)}$, which is almost always done.

So, a rational plan, the implementation of which $\mathrm{DM}_k^{(\Im+1)}$ can expect from $\mathrm{DM}_k^{(\Im)}\left(k=\overline{1,N_\Im}\right)$—increasing function $\hat{r}_k^{(\Im)}$.

Under these conditions (optimal in terms of $\mathrm{DM}_k^{(\Im+1)}$ and his plan) each of $\mathrm{DM}_k^{(\Im)}\left(k=\overline{1,N_\Im}\right)$ looking for his solution:

$$\max_{f^*}\left\{J_\Sigma^{(k)}\right\}=\max\left\{\sum_{j=1}^{m}C_{jk}\left(\hat{r}_k^{(\Im)}\right)f_{jk}^{*}\left(\hat{r}_k^{(\Im)}\right)-\frac{1}{2r_k}\sum_{j=1}^{n_{y\Im}}\left[f_{jk}^{*}\left(\hat{r}_k^{(\Im)}\right)\right]\right\}$$
$$=\max\hat{r}_k^{(\Im)}\frac{\Pi^{(\Im)}}{\hat{R}^{(\Im)}}\left[\sum_{j=1}^{m}C_{jk}-\frac{\hat{r}_k^{(\Im)}}{2r_k^{(\Im)}}\cdot\frac{\Pi^{(\Im)}}{\hat{R}^{(\Im)}}\right] \tag{6.43}$$

Then the income of such an element of the system as $\mathrm{DM}_k^{(\Im)}$ from (6.43) is determined not only by its estimates $\hat{r}_k^{(\Im)}$, but also estimates of other elements included in R. The highest return from (6.43) can also be found by the Lagrange method and is achieved when

$$\hat{r}_k^{(\Im)}=\begin{cases}r_k^{(\Im,M)}, if\ \sum_{j=1}^{n_{y\Im}}f_{jk}(\hat{r}^*)<Ц_л^*\\ r_k^{(\Im,m)}, if\ \sum x_{jk}\,(\hat{r}^*)>Ц_л^*\end{cases} \tag{6.44}$$

and from (6.42) it follows that:

$$\max_{\hat{r}_i^{(\Im)}}J_\Sigma^{(k)}\hat{r}_k^{(\Im)}=\max J_\Sigma^{(k)}\left(\hat{r}_k^{(\Im)*},\cdots,\hat{r}_{k-1}^{(\Im)*},\hat{r}_k^{(\Im)},\hat{r}_{k+1}^{(\Im)*}\cdots,\hat{r}_{N_\Im}^{(\Im)*}\right)$$
$$r_k^{(\Im,m)}\le\hat{r}_k^{(\Im)}\le r_k^{(\Im,M)} \tag{6.45}$$

For the synthesis of rational connections and relations between $\mathrm{DM}_k^{(\Im+1)}$ and $\mathrm{DM}_k^{(\Im)}$ and on $\Im$ level between $\mathrm{DM}_k^{(\Im)}\left(k=\overline{1,N_\Im}\right)$, consisting in the development of rational values of estimates $\hat{r}_k^{(\Im)}$ in conditions when $\mathrm{DM}_k^{(\Im+1)}$ has already solved its optimization problem (6.41), (6.42), it is obviously necessary to solve the game $N_\Im$—persons with known (given) boundaries of estimates $r_k^{(\Im,m)} \leq \hat{r}_k^{(\Im)} \leq r_k^{(\Im,M)}$ how to use strategies:

$$\max_{\hat{r}_i(\Im)} J_\Sigma^{(k)} \hat{r}_k^{(\Im)} = \max J_\Sigma^{(k)} \left(\hat{r}_k^{(\Im)^*}, \ldots, \hat{r}_{k-1}^{(\Im)^*}, \hat{r}_k^{(\Im)}, \hat{r}_{k+1}^{(\Im)^*} \cdots, \hat{r}_{N_\Im}^{(\Im)^*} \right)$$
$$r_k^{(\Im,m)} \leq \hat{r}_k^{(\Im)} \leq r_k^{(\Im,M)} \tag{6.46}$$

It can be shown that the optimality of plans (6.42), (6.44) is provided only in the conditions of existence of guarantees for sufficiently high prices:

$$Z^{(\Im)} = Z^{(1)} + \sum_{k=1}^{N_\Im} Z_k^{(2)} + \sum_{i=1}^{N_\Im} Z_k^{(3)} \tag{6.47}$$

In the found ratios for the optimal values of the task plan, the rationality of efficiency coefficients and price values is significantly influenced by the number of active performers-elements $\mathrm{DM}_k^{(\Im)}$—$N_\Im$.

Therefore, in order to give the obtained results (6.44)–(6.47) the form, it is necessary to determine the rational value $N_\Im = N_\Im^*$.

The rational structure of a typical two-level fan subsystem for a TPM found in this way, considered as an integral part of it, should be built on the basis of the principle of economic expediency in the relationship between the elements—$\mathrm{DM}_k^{(\Im)}$ and $\mathrm{DM}_k^{(\Im+1)}$ according to the condition of strict centralization and optimal planning (6.42), (6.44), prices (6.47) and efficiency coefficients (6.45).

It should be noted that the rational number of active executive elements in a typical two-level fan subsystem is determined based on their information characteristics. However, when taking into account economic indicators, this number can be adjusted upwards, based on the above results.

Indeed, if the costs Z_k^* from (6.36) to present in the form of time and invested funds attributed to the unit and current expenses for the implementation of the plan, then the total costs $\mathrm{DM}_k^{(\Im+1)}$ and $N_\Im$ elements—$\mathrm{DM}_k^{(\Im)}$ will make up:

$$Z^{(\Im)} = Z^{(1)} + \sum_{k=1}^{N_\Im} Z_k^{(2)} + \sum_{i=1}^{N_\Im} Z_k^{(3)} \tag{6.48}$$

where $Z^{(1)}$—the costs of technical equipment, capital construction and other equipment, i.e. the general part of the system-center $\mathrm{DM}_k^{(\Im+1)}$ and elements—$\mathrm{DM}_k^{(\Im)}$, assigned to a unit of time;

$Z^{(2)}$—k-th costs $\mathrm{DM}_k^{(\Im)}$, usually $Z_k = Z_{k\pm1} = Z_{k\pm2} = Z_{N_\Im}$ and in most cases it is true for the TPM:

$$\sum_{k=1}^{N_\Im} Z_k^{(2)} = N_\Im Z^{(2)}$$

- maintenance costs in case of non-fulfillment of the plan in the k-th element, i.e. the k-th $DM_k^{(\Im)}$, solving the problem; when $P^{(k)}$—the probability of delay or non-fulfillment of the j-th task, i.e.:

$$Z^{(\Im)} = Z^{(1)} + N_{\Im} Z^{(2)} + \sum_{k=1}^{N} \sum_{j=1}^{n_{y1}} \lambda_{jk}^{(\Im)} P_j^{(k)} C_{jk}^{(\Im)}$$
$$= Z^{(\Im 1)} + N_{\Im} Z^{(\Im^2)} \sum_{j=1}^{m} \lambda_j^{(\Im)} P_j^{(k)} C_j^{(\Im)} \tag{6.49}$$

For probability $P_j^{(k)}$ assuming that the total task flow has the parameter λ, there are no cases of imposing requirements on maintenance and service time—a random process with an exponential law with parameter v, a system of differential equations is known:

$$\begin{aligned} &\dot{P}(0)(t_0) = -\lambda P(0)(t) + \nu P(1)(t) \quad P(n) - \text{ probability of having } n \text{ requirements} \\ &\dot{P}(n)(t) = \lambda P(n-1)(t) - (\lambda + n\nu) P(n)(t) + (n+1)\nu P(n+1)(t)\ (1 \le n \le N) \\ &\dot{P}(n)(t) = \lambda P(n-1)(t) - (\lambda + n\nu) P(n)(t) + N\nu P(n+1)(t) \quad (n \ge N) \end{aligned} \tag{6.50}$$

In steady-state mode, this system is solvable with respect to probability P_j for each type of requirements:

$$P_j = \frac{\lambda_j}{N!}\alpha^n \left[N\nu\left(1-\frac{\alpha}{N}\right)^2\right]^{-1} \left(\begin{array}{l} \frac{\lambda}{n!}\alpha^N \left[N\nu\left(1-\frac{\alpha}{N}\right)^2\right]^{-1} \\ + \frac{1}{(N-1)!}\alpha^N \left(1-\frac{\alpha}{N}\right)^{-1} + \sum_{n=1}^{N-1} \frac{1}{(n-1)!}\alpha^N \end{array} \right)^{-1} \tag{6.51}$$

where $\alpha = \frac{\lambda}{\nu} \cdot; \cdot \lambda < N\nu$.

The solution of the problem of the optimal number of elements is now found from (6.49) from the stationarity conditions taking into account (6.51) under the condition $N^* > \alpha$:

$$\frac{dZ}{dN} = Z^{(2)} + \sum_{j=1}^{m} \lambda_j C_j \frac{dP_j}{dN} = 0 \tag{6.52}$$

The solution of Eq. (6.52) is usually found numerically. It is obvious, for example, that the higher the cost of equipment of an element of the simplest control circuit of the form (Fig. 3.3) in relation to the maintenance costs of unresolved or delayed tasks, the lower the value of the number N * from (6.52), and vice versa, if the costs of solving delayed tasks increase, the more justified the introduction of additional elements—$\mathrm{DM}_k^{(\Im)}$ and the value of the number $N_{\Im}^*$ in (6.48) increases.

It is clear that the solution is sought in integers.

So, the ratios (6.42), (6.44), (6.46), (6.47) and (6.52) allow us to model the main functional dependencies that determine the material and economic mechanism using the principle of taking into account the personal factor in it, concretizing to a certain extent the previously obtained results.

References

1. Kryzhanovsky GA, Kupin VV, Plyasovskikh AP (2008) Theory of transport systems. In: Kryzhanovsky GA (ed). GA University, St. Petersburg
2. Kryzhanovsky GA, Shashkin VV (2001) Management of transport systems. Part 3. Severnazvezda, St. Petersburg 224s
3. Larichev OI (2002) Theory and methods of decision-making, 2nd edn, reprint.idop. Logos, Moscow, 392s
4. Lukinsky V (2007) Models and methods of logistics theory. In: Lukinsky VS (ed.) Uchebnoe-posobie. Peter, St. Petersburg, 447s
5. Galaburda VG, Persianov VA, Timoshinidr AA (1996) Unified transport system: Studies for universities. In: Galaburdy VG (ed.). Transport, Moscow, 295s
6. Kryzhanovsky GA, Chernyakov MV (1992) Integration of aviation information transmission systems. Transport, Moscow, 296c
7. Borisov VV, Kruglov VV, Fedulov AS (2007) Fuzzy models and networks. Hotline – Telecom, Moscow, 284s

Chapter 7
Physical Modeling of Transport Processes—Simulation Modeling, Training Complexes

7.1 Simulation Modeling of Transport Processes

The high complexity of transport processes, the presence of ergatic elements in the structure of their functioning makes the study rely mainly on the results of modeling. The effectiveness of this approach, especially in the study of large-scale processes in large-scale HAS zones, such as, for example, zones of air traffic, navigation or water traffic on rivers, automobile traffic, is due to its relatively wide spread, especially in highly developed countries.

So, since the 50s of the last century, the development of centers with simulation modeling of transport processes of various types of transport in France, Sweden, Great Britain, USA, Austria, in our country has begun. One of the characteristic examples is the rapid emergence of simulation complexes for the organization of air traffic control. A significant factor in the feasibility of their creation was the possibility of accelerated comparison with real processes of simulating the development of a dynamic air situation with various variants of the division of airspace, flight routes, radio navigation and/or satellite flight support and other organizational and structural solutions. Such modeling complexes have been created in most developed countries and are successfully developing, for example, in Russia on the basis of the State Research Institute of Aviation Systems, which uses a research software package for simulation of organization processes and ATC.

At the same time, the combination of the functions of research centers with training centers is caused by the following reasons. Each of the centers uses very expensive equipment, and equipping each of them requires significant costs. At the same time, the greatest difficulty in modeling is the description of the activities of operators included in the simplest contour and other elements of the general structure of the ATC system. That is why, in models of ATC systems, the functions of dispatchers and pilots are often not modeled, and real operators are included in the contour of the modeling device, reproducing the activities of dispatchers and pilots.

The simulation device constructed in this way can be used for carrying out research work in the case when experienced trained operators are at the controls

© The Author(s), under exclusive license to Springer Nature Singapore Pte Ltd. 2023

G. A. Kryzhanovsky et al., *Modeling of Transportation Aviation Processes*, Springer Aerospace Technology, https://doi.org/10.1007/978-981-19-7607-0_7

of pilots and dispatchers, as well as for initial training of dispatchers. Typically, such real-time simulation devices form research simulators and include computers, pilot operator consoles, control consoles, video recording and playback equipment, data transmission and processing, as well as other auxiliary devices.

Research dispatch simulators are usually used in the study and solution of a whole set of tasks, the main of which are: analysis and study of aircraft flows in specific ATC zones; study of the effectiveness of the ATC system in a certain region; determination of rational structural characteristics of ATC zones; study of the effectiveness of the functioning of individual elements of ATC systems (including new ones being introduced) with the purpose of identifying bottlenecks and the rational distribution of functions between the elements of the system; training, retraining and improving the skills of ATC specialists; conducting a wide range of ergonomic studies; checking the effectiveness of interaction between dispatchers of adjacent management sectors; working out the operational technology of dispatchers of various points, and many others.

Modeling complexes usually include several simulation devices and research dispatch simulators and other equipment. This allows you to quickly switch from the study of one of the tasks to another or to conduct them in parallel, provided that the computer is fast enough and they work in time-sharing mode. An important role in the functioning of modeling complexes is played not only by their equipment, the most complete picture of which can be obtained from a comparison of computer data used in such centers, but also by the mathematical support of computers, i.e. a set of mathematical models of processes.

Modeling of ATC processes and creation of modeling complexes is one of the central directions in the automation of ATC processes. Modeling, roughly speaking, is used in two rather interrelated directions. The first involves modeling individual operations, tasks and processes or entire groups of them at various stages of the operation of the ATC system for the purpose of automation. The second direction is related to the modeling of processes in the ATC system for a comprehensive and in-depth study of them. Such modeling is carried out in modeling complexes. The interrelation of both directions is manifested in the fact that mathematical models of processes include a number of similar simplest models. Such models are formed from individual tasks and operations that together make up a certain process. In accordance with the principle of decomposition and models of individual processes, the system can be divided into a number of separate elements—tasks and operations. It is possible to name whole sets of identical models of such tasks and operations, from which more complex models of ATC processes are formed. So, there is a whole direction of ATC systems research—the modeling method. At the same time, it is necessary to introduce a number of assumptions and take into account the relationship of process models, tasks and operations to implement an integrated approach to the study of models. Only such an approach allows us to reasonably talk about the direction, specific location and methods of automating operations, tasks and entire processes of the ATC system at various stages of its functioning. It is in this way that it is possible to distribute functions between a person and a machine at a certain justified level of improvement of ATC processes.

It should be noted that the main difficulties encountered in modeling transport processes consist in the need to solve two problems, one of which is a significant variety of processes occurring in the transport system during their functioning, and the other is a weak knowledge of a number of processes, such as the decision-making process, etc. The diversity of the physical nature of processes leads to a variety of types of mathematical models of them. It would be a great courage to undertake the description of all, albeit finite, but a significant number of sets of elementary models of operations and tasks forming mathematical models of processes in the transport system. The problem of describing aggregates is associated with significant difficulties, the main part of which is of a fundamental nature and consists in the fact that mathematical models of various types must be implemented using computing devices of the same type, or close to each other, which requires uniformity of models.

In addition, some of the processes are not yet amenable to modeling or have mathematical models of poor quality (in terms of adequacy, simplicity, etc.). The needs of practice force us to intensively develop modeling methods, while seeking new and new approaches that take into account the versatility of process research, using new methods of description and new means of implementing models. Some elementary models obtained in this way and the results of their application turn out to be successful and are widely distributed. If we make a temporary breakdown of the period of functioning of the transport system into the stages of the organization of the transport space and the processes of functioning, then by giving the most common models and characteristic tasks for each of them, we can obtain, taking into account the reservations made above, a fairly complete picture of the current state of the issue of modeling transport processes.

So, for example, in the complex of simulation modeling of the air traffic management system in the State Research Institute of Aviation, predictive studies are carried out on the use of airspace in order to rationalize it, support decision-making to improve the structure of airspace by minimizing the number of points of occurrence of possible dangerous approaches of aircraft, as well as planning air traffic flows in areas of intensive flights to achieve standards bandwidth. All such and similar studies and the solution of problems of an applied nature arising in a variety of zones of the airspace of Russia are possible only in the presence of powerful computing complexes, developed software based on the formed mathematical software. All this testifies to the widest application of mathematical modeling methods to create such semi-natural physical models that simulate the functioning of real transport processes.

7.2 Modeling of Training Processes for Transport Specialists

The high level and pace of development of transport technology, as well as the entire complex of scientific and technical support of the transport system, significantly increase the amount of knowledge and the number of skills that a specialist in transport process management should possess. Meanwhile, the possibilities of mastering a certain amount of information and the ability to acquire skills in a given time are not unlimited. Therefore, the only possible way to meet the ever-increasing requirements for the level of training of transport specialists is to dramatically increase the efficiency of the very process of their training. The most well-known methods of increasing efficiency are methods based on the use of programmed learning and the widespread introduction of technical training tools [1, 2]. If we confine ourselves to the consideration of issues only of the professional training of HAS operators, then this path turns out to be inextricably linked with the improvement of simulator training. Therefore, starting from the second half of the 50s, in countries with significant traffic intensity, along with the development and complication of the HAS themselves, the use of training simulators is also expanding. At the same time, there is a constant tendency to increase the number of functions performed by the simulator. This includes the possibility of simulating a dynamic transport situation, including various structures of the transport space, modeling the movement of a significant number of vehicles of various classes, modeling the operation of technical means of transport, radio communications, interference, weather phenomena, interaction of the dispatcher with crews and other vehicles [3, 4]. Thus, a high degree of reliability of simulated conditions is achieved, which makes it possible to normalize the increase in the number of operations, exercises and tasks being worked out.

The use of the most modern computers in simulators allows us to raise the question not only about the imitation of conditions and the possibility of their differentiation and replication but also about the automation of the learning process itself. Here, it is meant in the general scheme of the learning process to transfer some of the functions of the teacher–instructor to assess the level of training, develop tactics and training programs (set and order of operations, frequency of their repetition, etc.) to the simulator, which includes a computer with a sufficiently capacious memory and significant performance. The necessity and expediency of such a transfer of a number of teacher–instructor functions to a machine is justified by the fact that otherwise it is difficult to fulfill the basic principles and conditions corresponding to the optimal learning process. This conclusion is a consequence of the following reasons.

In the existing system of simulator training, the instructor, as a rule, deals simultaneously with many trainees (from 5 to 30), depending on the type of simulator and cannot promptly manage the learning process for each of them. In this case, the principle of an individual approach is largely violated, since it is difficult for one instructor to choose his own impact for each trainee every time.

The second reason is due to the fact that almost all operations, exercises and tasks performed on the simulator, as a rule, are designed for the "average" trainee. Therefore, one part of them does not work at full strength on the training, is distracted and loses the newly formed connections in the formation of key supporting fragments of activity (KFA), which immediately affects the level of assimilation of the trained skills. The other part of the trainees, on the contrary, does not have time to form stable KFA, which also affects the level of acquired skills. As a result, both groups of trainees have fragile skills in performing operations.

The third reason is that the instructor does not have the opportunity to carry out systematic and stimulating control. This makes it difficult for the instructor to assess the actual state of the learning process for each of the trainees.

This practically excludes the instructor's analysis of the correctness of the learner's understanding of the goals, i.e. the identification of the magnitude $\Delta = z - z^*$—the degree of misunderstanding by the learner of the true meaning, for example, of the operation of the decision-making process.

If we assume that the tactical and technical data of the computer allow the implementation of sufficiently complex algorithms that optimize the learning process for each of the students, then the whole task of improving the efficiency of the learning process focuses mainly on the development of such algorithms. Roughly, the approximate scheme of the learning process when the computer is included in the circuit "teacher–instructor—trainee" is as follows.

A set of exercises and tasks interconnected in the form of a hierarchical system is stored in the memory of the computer (see Fig. 7.1). For each of the operations, each exercise and task, an indicator is known and entered into the machine that determines the acceptable level of quality of performing this operation, at which the training is considered completed. Differentiation of learning outcomes in the form of assessments is also acceptable here. According to the results of the actions of the trainee on the simulator, the state of the process of his individual training is assessed using the program. Comparing the state of the learning process with the indicator, the training program develops further effects—operations, exercises and tasks that are worked out in a certain amount.

The training program can serve as a variant of the machine implementation of solving problems of optimizing the learning process. In the first of them, when teaching operational skills, the optimal functions of direct control are selected by the machine, such as the training time for practicing this operation, the number of repetitions and others, as well as the sequence of operations, which determines the optimal functions of indirect control of the learning process. All these data are communicated via a special screen or scoreboard to the instructor conducting the training. Similarly, in the second task, optimal values of the degree of influence on the learner are developed. This in the "machine" version (where the computer acts together with the instructor) corresponds to the assessments and comments issued to the instructor by the machine on the light board or by the "voice system". Thus, automation of the learning process is possible in all variants, when the activity of the instructor is significantly facilitated.

The main difficulties in implementing such a scheme are associated with the development of computer algorithms that solve the problems of optimizing the learning process for each of the trainees, and the justification of indicators that determine the acceptable level of operations, exercises and tasks being worked out, especially for complicated variants.

If the development of algorithms for optimizing the learning process can be carried out using various kinds of simplifications, then it is not possible to fully automate the learning process without justification and formalization of learning indicators (determining the acceptable level of quality of each operation, exercise and task).

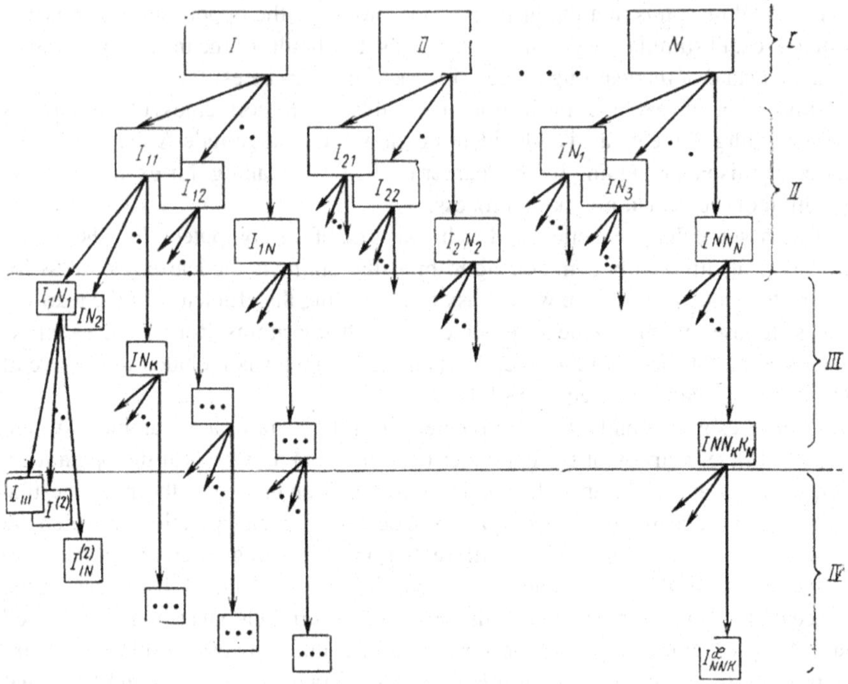

Fig. 7.1 Decomposition of the learning process: I—number of training stages N; II—the number of tasks worked out at all stages; $\underset{i=1}{N'} = \sum_{j=1}^{N_i} I_{ij}$; III—the number of exercises performed during training; $N = (2) \sum_{i=1}^{N} \sum_{j=1}^{N_N} \sum_{r=1}^{N_n^N} I_{ijr}$; IV—number of operations

Let us consider one of the possible variants of the method of constructing such indicators in the form of a comprehensive assessment of the student's activity in the process of working out operations, exercises and tasks. At the same time, the results obtained during the development of process efficiency indicators in the HAS can be actively used.

One of the possible approaches to the problem of constructing comprehensive assessments of trainees is based on the application of the decomposition principle, which allows us to identify a number of typical operations, exercises and decision-making tasks, each of which is characterized by its own local goals, mathematical models, methods of formalization and solution (see Fig. 7.1). The tasks identified in this way, interconnected by the technological process of the HAS functioning, allow us to build a hierarchical structure of performance indicators and assessments of activities, both of the trainee in the learning process and of the operator directly managing the transport process. At the same time, it is believed that the efficiency indicator $J_{\sum(v)}^{(r)}$ at a higher r-th hierarchical level, corresponding to enlarged tasks, it can be expressed as a functional dependence on a set of parameters and factors, where the latter are indicators of the effectiveness of solving exercises $J_{\sum}^{(r+1)}$ allocated at the underlying $(r + 1)$-th level, for example, in the form of a weighted sum:

$$J_{\Sigma}^{(r)} = \sum_{v=1}^{\mu} \omega_{v}^{(r)} I_{\Sigma(v)}^{(r+1)}$$

where $\omega_{v}^{(r)}$—parameters reflecting the significance (weight) of providing extreme properties when solving exercises of the $(r + 1)$-th level to achieve the necessary effectiveness of tasks at the r-th level.

In general, the solution to the problem of constructing such complex indicators that allow assessing the level of training and quality of activity is based on an analysis of the qualitative and quantitative characteristics of the process implemented in solving a typical problem with the management of the transport process, taking into account the features imposed on the HAS as a whole by its individual links. At the same time, it often turns out that the ratio of rationality (or presumably optimality) of the student's activity in the specific conditions of the problem being solved can be established only at a qualitative descriptive level. Therefore, the issues of formal description of the criteria of optimal (rational) activity of the student are relevant.

It is possible to achieve this goal with the help of an approach based on the principle of evaluating the final result of the real activity of an experienced operator. It is assumed that some experienced and competent operator carries out activities in an optimal (suboptimal) way in the sense of a certain optimality criterion, reflecting in an integral form the representations of this operator about the goals and required management results when solving the tasks facing him in a real situation. Such an assumption is valid if the solution of tasks in the transport process is carried out in compliance with the established regulations, which determine, explicitly or implicitly, the list of required goals, as well as the rules of decision-making.

In this case, the formalization of the optimality criterion of activity in the field of the extremum, which the operator operates during the control of the transport process, is possible with the help of solutions to the inverse problem of the theory of optimal processes, the essence of which is to establish the structure and parameters of the optimality criterion based on the known results of solutions to control problems that reflect the extreme properties of the operator's activity in specific conditions.

In general, in a real situation, instability (variability) of the criteria for the optimality of the operator's activity is possible, manifested in the participation of various operators and DM during the management of transport processes, non-stationarity of professional and psychological properties and qualities in transport processes. In these conditions, important tasks arise to assess the degree of decline in the quality of the system's functioning due to the manifestation of these factors, as well as to substantiate the adequacy and objectivity of the construction of criteria for optimal performance. The mentioned difficulties can be overcome by the joint use of methods of the theory of the inverse optimization problem and methods of expert evaluation for the construction of comprehensive assessments of activities. The specifics of the application of these methods will largely be determined by the nature of mathematical models of processes for solving problems of managing transport processes. The process of functioning of the HAS (in real conditions or to ensure the representativeness of samples on special simulators) is repeatedly observed by a group of experts, and is also subject to hardware registration.

For each i-th implementation, based on the values of the control function and the vector of generalized phase coordinates identified from the registration data (deviations from the program trajectory), the inverse problem of the theory of optimal processes is solved, i.e., the generalized characteristic of the *i*-th process realization is found, which determines the integral properties and qualities possessed in the process of a specific i-th implementation of technical and ergatic elements of the system. With constant technical characteristics of the system (the same type of vehicles, the use of the same means of displaying information, etc.), and also taking into account the positive motivation for activity, both the operator and the LPR, and the trainee, it can be assumed that the control process in each implementation is carried out by the operator using extreme capabilities, which are determined by mainly professional and psychological qualities. In this sense, the constructed generalized characteristic can be considered as an indicator of the optimality of the activity of the trainee and the operator with the appropriate level of professional training and readiness for activity in the conditions of the *i*-th implementation.

Each i-th implementation of the management process, in addition, is subjected to one of the peer review procedures. The analysis and processing of the results of the expert assessment allow the division of the indicators and their corresponding indicators into a number of classes in accordance with the preference system used in the expert assessment. The parameters forming the core of the optimality index in each class are averaged according to one of the smoothing procedures.

Classes, in turn, are characterized by different levels of quality of the system functioning and, consequently, are determined by different requirements for the level of quality corresponding to the situation of the class of tasks, and hence the entire professional training of trainees and operators on the development of their skills. The possibility of such justification of the construction and classification of indicators creates prerequisites for an objective assessment of the quality of the activities of trainees and operators at the most difficult stage, both in the learning process and in real conditions.

Using basic indicators J^S to assess the activity of the operator or the level of training of the trainee, it can be based on solving the inverse optimization problem with respect to the obtained optimality indicator *J*, which is attributed to one of the classes according to the scheme of determining the smallest distance *J* from the base $J^S(S = \overline{1, N})$ in the space of their parameters.

The theory of the inverse problem makes it possible to obtain constructive solutions most simply in the case when the control functions: can be represented as linear combinations of generalized coordinates of the system, the change of which is described by a system of linear differential equations.

The proposed approach to the construction and use of a comprehensive assessment of the activity of the trainee (or) and the operator with the selected mathematical apparatus for formalizing the procedures considered can be automated using a computer included in the contour of the simulator used in the process of training and advanced training.

7.3 Simulator Training in the System of Professional Training of Operators. The Problem of Choice

Simulator training is one of the central types of professional training of ITS operators, for any of them this is confirmed by the composition of initial training programs and retraining programs for ITS operators. A significant number of works have also been devoted to the study of the processes of professional training of operators on simulators. The predominant role of simulator training is also emphasized in the work on professional training of operators of transport systems of any type of transport. Professional training of operators cannot be imagined without a significant training component in the military sphere. The significance and role of simulators in the training of operators of high-risk facilities and systems is also evidenced by a fairly extensive list of standards adopted for their certification. The role of simulator training is constantly emphasized in ICAO, other international transport organizations and in publications abroad.

Numerous and multifaceted publications devoted to the issues of simulator training in varying degrees of completeness characterize, classify training devices and complexes according to the degree of their compliance with simulated parameters and traffic conditions, considerable attention is also paid to the method of increasing the degree of such compliance. Thus, in the vast majority of publications, there is always a question of assessing the quality of professional training processes on simulators. To date, it is clear that such an assessment is of a composite nature and the problem can clearly be attributed to the problem of multi-criteria choice. Indeed, among the indicators of the effectiveness of these processes, there are inevitably: the completeness of the list of key fragments of activity (KFA); the simulators worked out on this variant; the degree of proximity of the physical and functional movement models simulated by the simulator; the completeness and adequacy of the environment provided by the visualization system; flexibility in managing the process of professional training, including the possibility of stopping, returning, repeating changes in the situation and entering special cases during movement; automation of the assessment of the degree of training of the trainee to any of the selected KFA, processing of this data and providing it to the training instructor; availability of the possibility of modernization, consisting in the replacement of individual systems, blocks and modules of the simulator; assessment of the reliability of the simulator. The construction of a quantitative assessment characterizing the preference function necessary for a reasonable choice of the type of simulator or rather their complex of several is a task whose solution is clearly not necessary to consider completed. This statement is also supported by the fact that the effectiveness of the simulator training processes significantly depends on the methodological and organizational characteristics of the entire educational process. That is why most serious studies come to the conclusion that it is practically impossible to build a single composite reasonable assessment reflecting the completeness of the coverage of the list of KFA, adequacy of the conditions and characteristics of movement, technical simplicity, reliability, methodological points and economic indicators of simulators and/or their complexes in use for training ITS operators. Moreover, it is practically impossible to take into account such a delicate moment as the relationship of the organization of the educational process, the methods of its implementation on the simulator with the problem of evaluating the quality of the simulator. As a result, attempts to solve the problem of assessing the quality of a simulator or their complex, regardless of determining their functional place in the general process of professional training of operators and the relationship with theoretical and practical training, can lead to serious problems in the level of readiness of trainees for the process of real activity—for their certification. It is for these reasons that the professional level of instructors–teachers of training centers plays an essential role in the processes, organization of the entire educational process and in the selection of training devices and their complexes. Thus, taking into account all these factors in a single composite mathematical dependence is possible only at the explication level of modeling, when such a dependence has the form:

$$J_{\Sigma}^{(i)} = \sum_{j=1}^{m} \omega_j I_j^{(i)}, \left(i = \overline{1, N}, j = \overline{1, m}\right) \tag{7.1}$$

where ω_j—coefficients of importance (weight) of the j-th indicator of quality of efficiency;

$I_1^{(i)}$—indicator of completeness of the list of KFA in the i-th form of the simulator.

$$I_1^{(i)} = \frac{n_i}{n_{\max}} \tag{7.2}$$

where n_i—the number of KFA worked out on the i-th type of simulator;

$n_{\max}$—the largest possible number of KFA achieved in the world practice of simulator training of pilots;

$I_2^{(i)}$—an indicator of the degree of proximity of the motion model to its reality.

$$I_2^{(i)} = \frac{k_i}{K} \tag{7.3}$$

where k_i—the number of simulated motion characteristics, sound, visual, vibration and other characteristics;

K—the largest number of them;

$I_3^{(i)}$—flexibility in managing the process of simulator training.

$$I_3^{(i)} = \frac{r_i}{R} \tag{7.4}$$

where r_i—the number of variants of changes in the process of coaching in the form of stop, return, repeat, loss of special cases and situations during movement;

R—the largest number of them implemented in the simulator building;

$I_4^{(i)}$- the degree of automation of the training assessment, including measuring the results of the training to any of the selected KFA, processing them and providing them to the instructor.

$$I_4^{(i)} = F_4^{(i)}\left(\sum_{l=1}^{n_i} \frac{C_l}{\hat{\delta}_l^{(i)}};\ A;\ f\left(\delta_l^{(i)}\right);\ \Pi\right) \tag{7.5}$$

where $\hat{\delta}_l^{(i)}$—evaluation of the measurement error of the results of the l-th KFA coaching $l = \overline{1, n_i}$; C_l—the importance coefficient of the l-th error; A—linguistic variable of the automation level; $f\left(\delta_l^{(i)}\right)$—a variant of processing measurement results to obtain its assessment $\hat{\delta}_l^{(i)}$; Π—a linguistic variable that characterizes the presentation of information to the instructor.

$I_5^{(i)}$—the degree of availability of the possibility of upgrading the i-th version of the simulator–linguistic assessment.

$$I_5^{(i)} = F_5^{(i)}(V_i) \tag{7.6}$$

$I_6^{(i)}$—the reliability assessment of the i-th version of the simulator can be obtained from information about the reliability of its elements; its normalized value is obtained by estimating the largest and known reliability values.

$$I_6^{(i)} = \prod_{k=1} \frac{P_k^{(i)}}{P_{\max}} \tag{7.7}$$

$I_7^{(i)}$—the cost, similarly, is estimated as a ratio.

$$I_7^{(i)} = \frac{S_{\min}}{S^{(i)}} \tag{7.8}$$

where $S_{\min}$—the lowest of the known prices of simulators of a similar type;

$S^{(i)}$—the cost of the i-th simulator;

$I_8^{(i)}$—linguistic assessment of the compliance of this i-th type of simulator with methodological requirements.

$$I_8^{(i)} = F_8(M_i) \tag{7.9}$$

Thus, if we introduce quantitative estimates of linguistic variables and estimates in performance indicators $I_4^{(i)}$, $I_5^{(i)}$ and $I_8^{(i)}$, what is that when assigning importance coefficients as one of the methods of processing the results of the survey of expert instructors, the formula (7.1) can provide significant assistance in choosing the type of simulator.

The presence of performance indicators having a qualitative–linguistic characteristic relates the problem of choosing a simulator option to poorly structured tasks. Decision-making in the form of choosing a simulator variant in a weakly structured problem (7.1–7.9) can be approximated by the methods recommended in the study of such processes in conditions of fuzzy sets. Indeed, linguistic assessments in indicators $I_4^{(i)}$, $I_5^{(i)}$ and $I_8^{(i)}$ are attributed to the fuzzy ones. Of the greatest interest is a group of decision-making methods used for the case with a small number of alternatives, typical for the task of choosing a simulator variant, one of the most currently used methods from this group is the method of analytical hierarchy by *T*. Saati. For its application, the so-called scale of quality gradation values is introduced, which allows the linguistic assessment of quality characteristics to be translated into a quantitative equivalent:

Now the classification values of indicators $I_4^{(i)}$, $I_5^{(i)}$ and $I_8^{(i)}$ can be estimated using quantitative equivalents from the Table 7.1.

So, having obtained with the help of dependencies (7.1–7.9) and Table 7.1, the quantitative values of all the listed performance indicators and by means of an expert procedure for evaluating the values of the importance coefficients ω_j.

You can choose from N variants of simulators the one that meets the requirement:

$$J_{\Sigma}^{*} = \max_{i} \sum_{j=1}^{m} \omega_j I_j^{(i)} \tag{7.10}$$

The practical activity of simulator training for more than ten years, however, indicates the low popularity of the seemingly theoretically justified quantitative approach

Table 7.1 Classification values of indicators

No.	Classification value of the indicator	Quantitative equivalent
1	Absolutely low	0.0
2	Very low	0.1
3	Low	0.2
4	Low enough	0.3
5	Medium	0.5
6	High enough	0.7
7	High	0.8
8	Very high	0.9
9	Absolutely high	1.0

described above. And it does not at all lie in the complexity of constructing mathematical dependencies (7.1–7.10). This approach is justified when carrying out export purchases of training equipment or a large-scale order for one of the selected options for its manufacture. In the real practice of the activity of training centers of educational institutions of the training profile of our country, and most of the foreign ones, these cases are quite rare. From a practical point of view, an approach seems more justified when, on the basis of clearly formed requirements for the properties of the simulator or their complex, a variant of its layout from already existing individual components is justified or their integration is carried out. It is this approach that makes it possible to obtain estimates, including quantitative ones, not only of the indicators listed above, but also to achieve some new qualitative result—the ability of a new simulator device to solve the tasks of training operators of a set of such KFA, the inculcation of which during simulator training on traditional simulators was unthinkable.

7.4 Model of a Variant of an Integrated Intelligent Training Device

The development of information technology, the global satellite navigation system and the introduction of all these innovations into transport processes required the development of new knowledge, skills and abilities from writers and DM who manage them, a number of which become key for a certain type of transport. Of particular importance is the expanding capabilities of training equipment. So there is an opportunity to expand the list of skills formed by air transport operators—pilots.

The formation of the required pilot orientation skills in a complex DAS when flying in various sectors of ATC is based on previously acquired knowledge and skills in piloting, navigation, conducting radio exchange according to rules and phraseology and other applied disciplines. However, the main method of forming the required skills in this case, as a rule, is in real or simulated conditions. Simulated conditions are naturally preferable due to a number of reasons, the main of which are ensuring the safety of professional training, its efficiency and a high degree of adequacy to real flight conditions.

To achieve such a high degree of adequacy of the DAS simulation, a comprehensive development of organizational, methodological, mathematical and technical support is necessary. The ideal model of such an integrated intelligent simulator device (IISD) should include a group of methodologists who develop and implement a briefing scenario, a corresponding flight scheme in all sectors of the ATC with the presence of certain special cases, the detection of which by listening to radio exchange and taking into account the briefing data, should be carried out by the trainee. The same group develops or uses ready-made versions of tests and processes the results of a survey of instructors to assess the level of professional training of trained pilots. In addition to this group, groups of instructors, pilot operators and dispatchers of all ATC points participate in the process of simulator flights. The organization of coordinated activities of these groups, as well as engineering and maintenance personnel, is the essence of organizational and methodological support. Mathematical support involves the preliminary development of flight schemes that implement the selected scenario and the solution of ATC tasks with special cases of movement in the presence or prerequisites for potential conflict situations. Technical and engineering support includes synchronization and integration of such training devices, such as, for example, video recording of a briefing, the process of documenting test results, carrying out a simulator flight according to the selected scenario and fixing the parameters of the SFA of the trainees during the flight, as well as their processing.

Thus, a schematic diagram of the IISD appears in the form (Fig. 7.2).

Its technical implementation is possible in various variants, starting with the simplest simulators on the screens of monitors and PCs and up to the use of screens and video recordings of the briefing and real models of cabins of currently operated aircraft such as Tu-154, B-737, A-320, etc.—Yak-18 T, Il-103, etc. together with a full a set of dispatching simulators of all sectors of the ATC to fully simulate the flight process in each of the variants of the schemes developed by a group of teaching methodologists. It is important to assess the sufficient degree of adequacy of each of the possible options. Such an assessment can be implemented on the basis of comparing the magnitude of the difference between the information model of the situation created using this variant of training devices and the information model of the real situation arising in each scheme. Comparing this value with the maximum permissible values (norms), it is possible to draw a conclusion about the

Thus, a schematic diagram of the IISD appears in the form (Fig. 7.2).

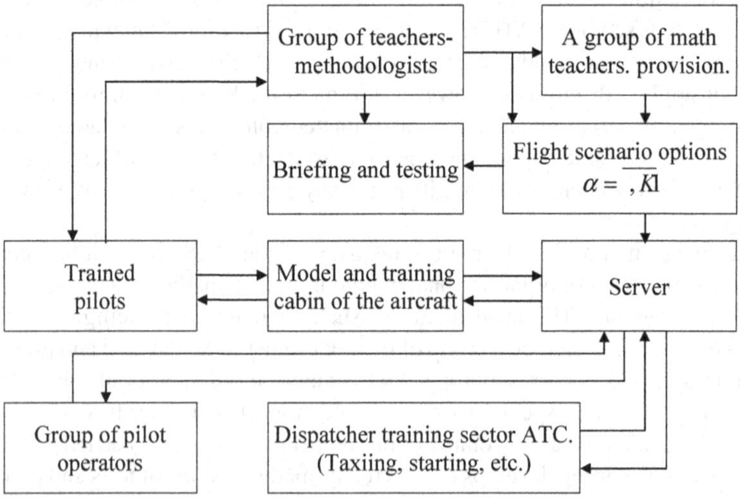

Fig. 7.2 Schematic diagram of IISD

acceptability of this option. It is important to take into account the adequacy of all the main types of information models—visual, and especially sound, which has the greatest importance for the formation of the required skills. Such accounting can be implemented by various methods. The simplest method is based on the theory of fuzzy sets and processing the results of a survey of a group of experts. A group of experts to assess the adequacy of information models of the visual and audio appearance of the simulator device (Fig. 7.2.) and the flight operation that actually exists in production conditions should be organized from among the commanders, co-pilots who have a significant flight hours on this type of aircraft, the cabin of which is used for simulation during simulator flights. The numerical composition of such a group should include at least 20–30 people with a flight of at least _____ hours of the category of persons of the I Class, and persons of the I class. Expert groups in the absence of the number of persons of the required quality of professionals can be created in stages—for example, in the process of passing advanced training courses by various categories of flight personnel. The results of the survey of each of the small groups are then taken into account in the general table, which is already the "total" result of the survey of the "virtual" large group. Each survey is carried out according to a single methodology after conducting a simulator flight according to one of the proposed scenarios with one of the proposed scenarios using a fuzzy scale of assessments forming a set of linguistic variable "adequacy". At the same time, each expert should assess the adequacy of the visual and audio information model by putting one or zero in the range of possibilities of the scale in which he considers this degree of adequacy to be the most appropriate. Intervals are formed from 0 to 100% of possible adequacy, which, in turn, is evaluated on a 7-point scale:

7—complete coincidence, complete adequacy without differences;
6—very large coincidence with minor deviations;
5—there are more matches than deviations;
4—almost equal number of coincidences and deviations;
3—there are matches with a larger number of deviations;
2—there are some similarities, but the deviations are significant;
1—lack of coincidences in information models.

As a result of summing up the recorded opinions of experts, the frequencies of their use of one or another degree of adequacy are formed (Table 7.2).:

Tables of the form (7.2) are filled in by a group of experts for two cases of adequacy assessment—visual and audio. In both cases, such data require preliminary processing to reduce the influence of erroneous opinions, which are characterized, for example, by the presence of a number of zeros (2–3) in the row of the selected element, for example, two experts put one in the 3rd row of the fourth interval, whereas the I-th, II-th, III-th, V-th, VI-th contain only zeros:

$x_{3,1} = 0; x_{3,2} = 0; x_{3,3} = 0; x_{3,4} = 2; x_{3,5} = 0; x_{3,6} = 0$; etc.

Then the value $x_{3,4} = 0$ it is corrected to zero and a so-called summary matrix is constructed, which contains only one row—the sum of all elements by columns:

Table 7.2 Assessment of the adequacy of the information model

Intervals degree of adequacy	1	2	3	4	5	6	7	8	9	10
	0–10	11–20	21–30	31–40	41–50	51–60	61–70	71–80	81–90	91–100
7	$x_{7,1}$	$x_{7,2}$	$x_{7,3}$	$x_{7,4}$	$x_{7,5}$	$x_{7,6}$	$x_{7,7}$	$x_{7,1}$	$x_{7,8}$	$x_{7,10}$
6	$x_{6,1}$	$x_{6,2}$	$x_{6,3}$	$x_{6,4}$	$x_{6,5}$	$x_{6,6}$	$x_{6,7}$	$x_{6,1}$	$x_{6,8}$	$x_{6,10}$
5	$x_{5,1}$	$x_{5,2}$	$x_{5,3}$	$x_{5,4}$	$x_{5,5}$	$x_{5,6}$	$x_{5,7}$	$x_{5,1}$	$x_{5,8}$	$x_{5,10}$
4	$x_{4,1}$	$x_{4,2}$	$x_{4,3}$	$x_{4,4}$	$x_{4,5}$	$x_{4,6}$	$x_{4,7}$	$x_{4,1}$	$x_{4,8}$	$x_{4,10}$
3	$x_{3,1}$	$x_{3,2}$	$x_{3,3}$	$x_{3,4}$	$x_{3,5}$	$x_{3,6}$	$x_{3,7}$	$x_{3,1}$	$x_{3,8}$	$x_{3,10}$
2	$x_{2,1}$	$x_{2,2}$	$x_{2,3}$	$x_{2,4}$	$x_{2,5}$	$x_{2,6}$	$x_{2,7}$	$x_{2,1}$	$x_{2,8}$	$x_{2,10}$
1	$x_{1,1}$	$x_{1,2}$	$x_{1,3}$	$x_{1,4}$	$x_{1,5}$	$x_{1,6}$	$x_{1,7}$	$x_{1,1}$	$x_{1,8}$	$x_{1,10}$

$$x_j = \sum_{j=1}^{7} x_{ij};\ \left(i = \overline{1, 10}\right) \tag{7.11}$$

where x_{ij}—the sum of expert assessments on the i-th degree of adequacy of the probability values falling on the j-th interval.

These operations are performed for two frequency tables—visual and audio, and as a result, two total matrix rows are obtained $\overline{x}_i^{(b)}$ and $\overline{x}_i^{(3)}$ kind of (7.11).

Next, the corrected matrices are constructed $Y^{(b)}$, $Y^{(3)}$ expert opinions on the assessment of two degrees of adequacy, respectively, according to the dependencies:

$$\left.\begin{aligned} y_{ij}^{(b)} &= x_{ij}^{(b)} - \frac{\overline{x}_{\max}^{(b)}}{\overline{x}_j^{(b)}} \\ y_{ij}^{(3)} &= x_{ij}^{(3)} - \frac{\overline{x}_{\max}^{(3)}}{\overline{x}_j^{(3)}} \end{aligned}\right\} \tag{7.12}$$

where $\overline{x}_{\max}^{(b)} = \max\limits_j\left(\overline{x}_j^{(b)}\right),\quad \overline{x}_{\max}^{(3)} = \max\limits_j\left(\overline{x}_j^{(3)}\right),\quad \left(j = \overline{1, 10}\right)$

Finally, according to the data of the matrices $Y^{(b)}$ and $Y^{(3)}$, you can get the type of membership functions:

$$\mu^{(b)}\left(y^{(b)}\right) \text{ and } \mu^{(3)}\left(y^{(b)}\right)$$

finding their points on the probability axis:

$$\left.\begin{aligned} \mu_{ij}^{(b)} &= \frac{y_{ij}^{(b)}}{y_{i\max}^{(b)}} \\ \mu_{ij}^{(3)} &= \frac{y_{ij}^{(3)}}{y_{i\max}^{(3)}} \end{aligned}\right\} \tag{7.13}$$

where $y_{i\max}^{(b)} = \max_i\left(y_{ij}^{(b)}\right), \quad y_{i\max}^{(3)} = \max_i\left(y_{ij}^{(3)}\right), \quad \left(i = \overline{1,7};\ j = \overline{1,10}\right)$

Having thus constructed the functions of belonging to the linguistic variable "adequacy" to assess the degree of visual and audio similarity of information flows in the simulator device and in real flight operation conditions, it is possible to draw reasonable conclusions about the quality of imitation and the degree of their similarity.

References

1. Kryzhanovsky GA, Kupin VV, Plyasovskikh AP (2008) Theory of transport systems, Kryzhanovsky GA (ed). GA University, St. Petersburg
2. Kovalenko GV, Kryzhanovsky GA, Sukhoi HH, Khoroshavtsev YE (1996) Improvement of professional training of flight and dispatching personnel, Kryzhanovsky GAM (ed). Transport
3. Nedzelsky II (2002) Marine navigation simulators: problems of choice.GNCRF-Central Research Institute "Electropribor", St. Petersburg, 220s
4. Damin LS, Zhukovsky YG, Semeniv AP et al (1986) Automated training systems for professional training of aircraft operators, Shukshunov BE (ed). Mechanical Engineering, 240s

Chapter 8
Modeling of Elements Characterizing the Activities of Operators and DM of Transport Processes

8.1 Motivation and Volitional Tendencies of Operators and DM of Transport Processes

In order to understand the reasons for the actions, actions and all activities of operators and DM of active transport systems (ATS), it is necessary to reveal the driving force, intentions, and what motivates them to these actions. At the same time, such motives are considered as time–stable phenomena and are called motives. Having divided all the motives into those that are caused by physiological needs and those that are related to the goals of achieving professional skill as meaningful areas of action, the latter are determined as the subject of research when analyzing the processes of professional training of operators and DM ATS. The study of this type of motives has an extensive history [1–3], which led to the development of a number of special methods for measuring the motive [4].

The generalizing concept of various processes and phenomena that determine actions, actions and behavior, taking into account the awareness of the consequences and their direction, proportionality with the expenditure of energy and volitional efforts, is the concept of motivation. Such processes and phenomena that form motivation represent a preparatory stage for the decision-making processes of the DM and for the implementation of actions in the form of speech-functional acts (SFA) that directly affect changes in the road traffic situation (DTS). The resulting state parameters are just the final indicators of the level of professional skill of this operator or DM ATS.

The complexity of motivation research also lies in the fact that, as a rule, the emerging states of DTS, more precisely, the perceived information about them, causes several motivations, each of which has its own goal and strives (encourages) to achieve it, which is called the final motivational trend. In a typical situation of the DTS state, there are several motivational tendencies for any operator and LDPR at the same time. But only one of them becomes the dominant and defining DMP and SFA. The processes that determine the choice of the dominant motivational tendency are called volitional processes [4].

© The Author(s), under exclusive license to Springer Nature Singapore Pte Ltd. 2023

G. A. Kryzhanovsky et al., *Modeling of Transportation Aviation Processes*, Springer Aerospace Technology, https://doi.org/10.1007/978-981-19-7607-0_8

The dominant motivational tendency forms something similar to an action plan. This plan is usually not a clear list of operations and their sequence. It should rather be considered an intention, which is included in the concept of the dominant motivation—tendency. It can also include such volitional processes that are carried out by the DMP when choosing from a number of intentions one that is implemented in the DMP and the SFA. Thus, it can be concluded that the information perceived by the operator and /or the DM determines a number of motivational tendencies corresponding to the situation of the DTS, the inclusion of volitional processes determines the dominant motivational tendencies—the choice of intentions and, by connecting memory (experience, professional level, etc.), makes a choice—the DMP about the type of SFA.

Motivation influences memory and the intensity of volitional processes. The relationship between the processes of perception of information about DTS and the formation of SFA can be conditionally represented using Fig. 8.1. The choice (DMP) about the type of SFA can be represented as a comparison of the expected (predicted) by the operator and/or DM development of the state of DTS and evaluation (comparison) of possible (probable) undesirable (dangerous) states. At the same time, it is necessary to take into account the constant factor of lack of time for DMP at ATS operators and the presence, as a consequence, of other influences of errors in the perception of information about the DTS and during its processing. Motivation, as a preparatory stage for DMP, generates a state of doubt in the mind of the operator and/or DM, which in turn generates a number of possible motivational tendencies, each of which is focused on the realization of its desired state of DTS and should take into account possible consequences. It is for this reason that it is possible to assume the type of efficiency indicator of this ith operator and/or DM ATS, one way or another reflecting the state of the DTS, their forecast (risks) of dangerous phenomena and the effort—costs that, according to his assumption, the implementation of each of the desired states will require, taking into account errors.

An attempt to formalize the preference function (PF) as a function determining the semantic-volitional process (SVP-I) of choosing from a number of motivational tendencies the one that becomes dominant and allows unambiguously implementing SVP-II–DMP–SFA.

Mathematical modeling of the choice in accordance with SVP-I can be carried out under the assumption of the expediency and rational activity of the operator and/or DM ATS, which corresponds to its certification as a specialist with the necessary level of professional training and health status (including the psyche).

Let a given ith operator and/or DM have a set $X^{(S)}$ motivational trends. The essence of the problem of modeling PF is to find the type of PF $U(X)$, such that:

$$\max_{X} J_{\Sigma}^{(S)} \to U^{(S)}(X),\ x \in X^{(S)} \text{ finite set} \tag{8.1}$$

The difficulty of determining the PF is due to the very nature of the physical essence of the operator and/or the operator of the DM ATS fixed at the "output" by a number of parameters of actions of activity and all behavior that usually have random

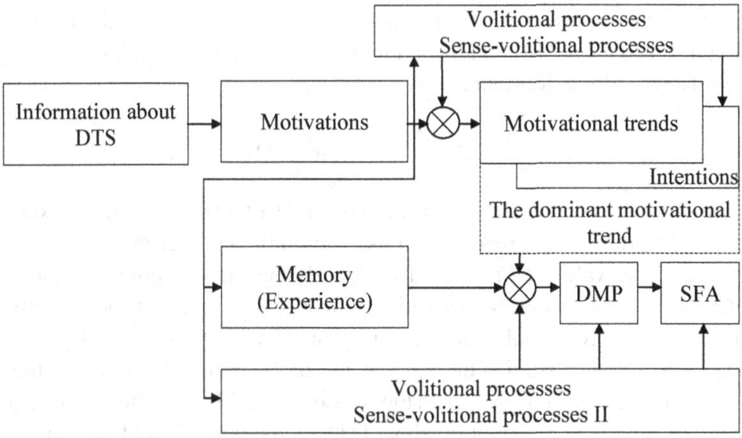

Fig. 8.1 Algorithm of volitional processes in DMP

processes and possible characters. This is also clear from the functional diagrams (Fig. 3.7), where it is shown that in the formation of activity, emotional coloring and subjective perception of information about DTS play an essential role. It is for this reason that most of the studies of the PF make the basic assumption that there is a stable attitude of the operator and/or the DM to the results of their activities (self-assessment) for a sufficiently long period of time, i.e. the existence of a stable limited finite set $X^{(S)}$ and the possibility of ordering its elements from the point of view of the operator and/or DM ATS [1–6]. The utility theory that has emerged from these and other studies unequivocally asserts the possibility of constructing an PF, with the help of which the choice of SVD-I is carried out. Here, of course, it should be taken into account that such a choice of one dominant motivational tendency in this particular situation always bears the imprint of the general orientation of the operator and/ or DM—his handwriting, thereby determining the features of all his activities and further behavior: SFA—act the totality of the PF activity total activity—behavior (line of conduct) [4–6].

If you define the set roughly enough $X^{(S)}$ consisting of motivational tendencies characteristic of operators and/or DM ATS, for example, of the form $x_1^{(S)}$—economic preferences; $x_2^{(S)}$—the desire to avoid difficult situations in DTS—fear of risk; $x_3^{(S)}$—the desire to gain leadership in the team; $x_4^{(S)}$—career aspirations, etc., then it is clear that the choice of the dominant one will make it possible to determine not one value of the indicator (1), but the whole area.

Motivations are formed by so-called quasi-needs and can be represented by motivational tendencies, from which a number of resultant (not one!) are selected, which form plans—intentions, of which volitional is called the dominant motivation. At the same time, the intensity of such a dominant motivation obeys the principle of the necessary costs for the implementation of the SFA: when difficulties arise, it increases motivational arousal—up to a certain limit determined by this dominant motivation, as well as the amount of energy and volitional costs, and the duration of their action. It is believed [4] that the strength of volitional tendencies can be equated with the strength of the dominant motivation

$$F_{\Sigma}^{(s)}(t) \approx k^{(s)} \cdot W^{(s)}(t) \tag{8.2}$$

It should be noted that there is a known effect of overmotivation (exam, etc.) leading to a deterioration in results—an increase in the number of errors $P_{\text{ош}}$ and/or an increase in the value $\tau_{\text{зат}}$ in the case of DMP and their implementation in the form of SFA. The effect of low motivation power is manifested in the low efficiency of achieving the professional-thinking ability of PTA in the process of professional training. Little has been studied on the question of the mutual influence of the intensity of volitional efforts and DMP (and the actions of the SFA) on the one hand and the processing of perceived information about DTS on the other. Most likely, in Eq. (8.2) on the right, it is necessary to add terms characterizing changes in volitional efforts regulating (restraining or intensifying) DMP and SFA and arising when compared with the intentions of the emotions accompanying them. From the moment the goal is achieved—the evaluation of the result of the SFA, as a result of volitional processes that implement dominant motivational intentions, a new systemquant begins—intentions, action plans, the expected result and their consequences, compared with the estimated state of the DTS and the costs incurred (energy and volitional)—this is how the memory chain is formed, and as a result—the acquired experience of mastering a set of key fragments of the KFA activities.

$$F_{\Sigma}^{(s)} \approx k\left(W_0^{(s)}(t) - \Delta W^{(s)}(t)\right) \rightarrow I^{(s)}\left[DTS(t_{k+1}) - DTS(t_k)\right] \rightarrow \pm \text{Э}^{(s)}(t_{k+1}), \tag{8.3}$$

that is, the self-esteem element of emotions associated with the desire to achieve either satisfaction (pride!)—and/or to avoid shame—extreme dissatisfaction. It is the result of such self-assessment that serves as the beginning of a new systemquant.

Thus, the most important elements of the functional education system of the Russian Federation are the presence of classes of values that select the most important and permissible needs for actions and decisions at an early stage. Then motives and intentions are formed—intentions. In accordance with the perceived information about DTS, these intentions are realized in the form of motivational excitations, which, in accordance with the accumulated experience—memory, emotions and volitional efforts, form the dominant motivation. For its implementation, volitional efforts are intensified, striving to achieve the goal through the choice and DMP of the SFA. At the same time, memory—experience is usually included and the DM shows its handwriting in the choice of DMP and SFA. Thus, two time periods can be distinguished—the period of the emergence of motivational arousals and educated intentions in the form of a dominant and the period of volitional processes—the initiation of the choice of DMP and SFA and tracking—bringing the SFA to the realization of the dominant. This is followed by the motivation of the aftereffect of evaluating the effectiveness of the SFA and the emergence of emotional self-esteem.

At the same time, the force of volitional influence corresponding to the requirements of the situation changed the DTS (the occurrence of difficulties) Eq. (8.3). All these processes serve as the basis for gaining experience—memory and strengthening PTA.

The values of the strength of volitional tendencies in Eqs. (8.2) and (8.3) are caused by the strength of the dominant motivational tendency. It also determines—selects from memory—experience and those decisions—DMP, which are then implemented (under the control of volitional processes) in the SFA.

However, this realization occurs as a result of conscious (or not so conscious) answers to a number of questions (available answers in memory—experience) that form a kind of model of DMP, the essence of which is to assess the subjective probability—the possibility of achieving the goal—satisfaction of the dominant motivational tendency (see Fig. 8.2).

The process of self-assessment also plays an essential role in this: the comparison of the achieved result (based on the perceived information about the state of the DTS) and the image of the situation that is attributed by the operator and/or the DM to a typical, complicated, complex or extremely complex. This is how an emotional assessment arises based on existing (formed) claims and incentives for further actions with the indispensable participation of volitional processes with the values of volitional tendencies determined by the strength of the dominant motivational tendency.

Repeated repetition of such systemquants makes it possible to achieve significant success in improving professional skills.

8.2 Intellectual Activity of ATS Operators

The practice of analyzing the activities of operators of hierarchical active transport systems (HATS) (pilots, commanders of sea, river vessels, air traffic controllers, etc.) shows that a well-developed apparatus for mathematical preparation and decision-making using the most advanced computing systems is most likely applicable only at the stages of their professional formation and, less often, in the processes of pre-trip analysis. Therefore, the conclusion about the general validity of models of the mathematical theory of decision-making in the activity of HATS operators can be considered premature so far. The fact is that decision-making processes in such complex ergatic systems, which are transport systems today and in the near future, are carried out mainly by a human operator, less often with the help of automated systems based on computer systems equipped with algorithms for optimizing decision-making processes. Therefore, analyzing the activity of operators of this type, one cannot but agree with the fact that in the process of their decision-making, first of all, a number of phenomena that characterize only for ergatic elements should be taken into account. This is a complex of emotional and volitional influences and other psychological factors and, of course, the impact of his (operator's) organization intellectual activity, which is the essence of decision-making. It is known that the decision-making procedure as an intellectual activity almost everywhere

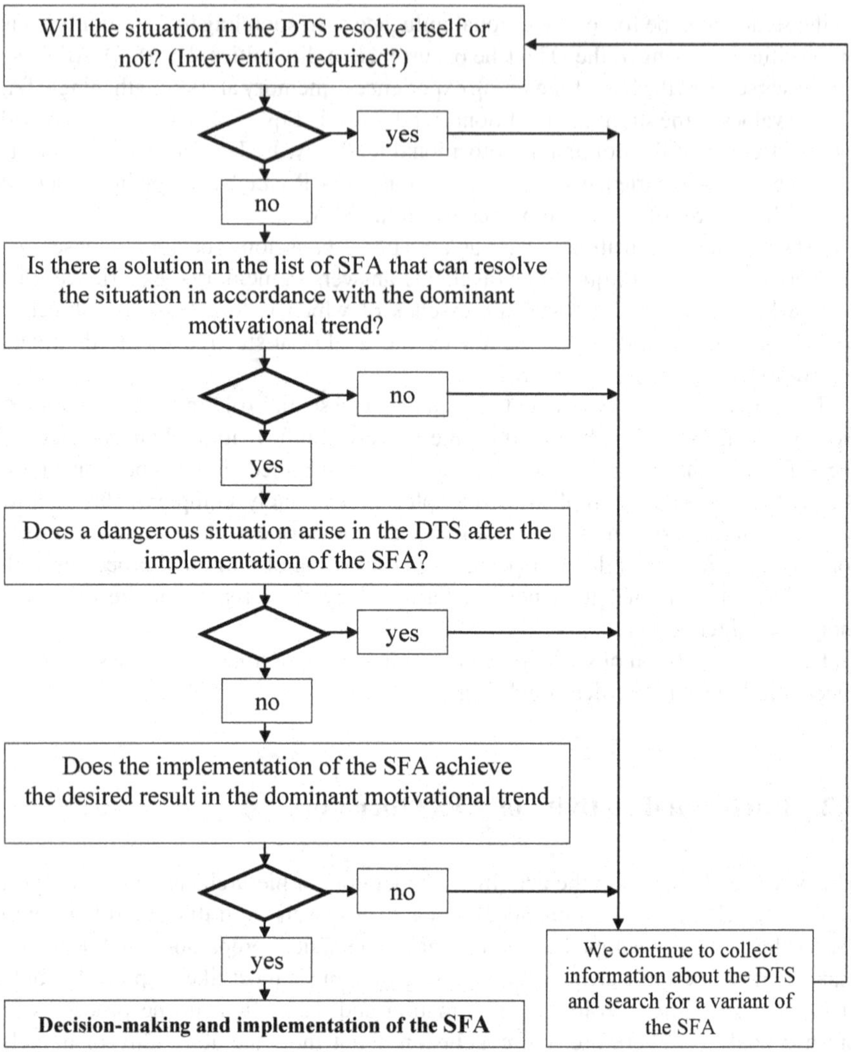

Fig. 8.2 Algorithm of decision-making and implementation of the SFA

has the same foundations, schemes and methodological algorithm. The bases of such an approach to the problems of human decision-making were laid taking into account psycho-physiological aspects—and taking into account the organization of intellectual activity—in the works [5–7].

It is advisable, when modeling transport processes, to attempt to discuss the ideas put forward in these and other studies that determine the intellectual activity of ATS operators in the decision-making process without their support from automated systems. The purpose of this discussion is an attempt to model decision-making processes. It is clear that the development of such models will serve the further development of intelligent active transport systems by improving their artificial intelligence subsystems.

The analysis of the structure and interactions between the blocks of functional systems for the formation of speech-functional acts (SFA)—the implementation of decision-making processes (DMP) by PBX operators clearly shows the role and importance of professional thinking ability (PTA) in such processes. For the development of PTA as an intellectual activity, critical thinking is of particular importance, aimed at identifying the structural features of reasoning, checking the correspondence between the operator's awareness of the state of the DTS and fragments extracted from memory (or by illumination), assessing the consequences of using the data of the DMP—SFA, rechecking the course of reasoning in the process of awareness (perception of information during the whole and totality (visual, auditory, etc.) of the state of the DTS, the use of memory and the assessment of consequences—and all this in conditions of, as a rule, a shortage of time. It is clear that the skill of critical thinking is one of the most urgently needed characteristics for a ATS operator in his quest to achieve high professional qualities. Critical thinking includes as the main characteristic—the skill of reflection on one's own mental activity, the properties of the analytical way of thinking, the ability to correlate concepts (about the states of DTS) and assess their characteristics, possible outcomes and their frequency—probabilities, to identify key issues that may most accurately and briefly determine the information necessary for the formation of DMP. Critical thinking includes the analysis of practical, professionally oriented reasoning (mental, often deciphered if necessary in the process of analysis), always bearing the personal characteristics of the operator. Critical thinking is inseparable from what can be called a struggle of opinions—the presence of several answers—solutions to each of the identified key issues. Here, the most important element in comparing answers can be the use of logic, including the results of mathematical logic. Logic makes it possible, while maintaining stability and consistency in the chain of practical reasoning—conclusions and the use of memory—knowledge, to take a critical approach to alternative options of the DMP—SFA. Thus, critical thinking, as one of the forms of rational thinking, organizes and brings harmony to the variety of images and fantasies that arise when searching for an acceptable DMP—SFA. It sets samples of ways of logical construction of reasoning and argumentation, being a certain methodology and style of rational thinking, and a necessary element of the formation of decision-making processes in general, as a form of intellectual activity of the ATS operator, testifies to the established PTA and the level/volume of the knowledge base formed, and in the case of rational decisions in non-standard complex situations when the corresponding states of the DTS using non–instructive innovative solutions—on the development of new elements of the knowledge base of this *s*th operator.

One of the characteristic elements of critical thinking is a balanced (in conditions of time shortage) delay in judgments. Such a delay is necessary to evaluate all the proposed solutions for establishing the most accurate list of questions and the sequence of answers leading to the justification and argumentation of a rational DMP. The decision is formed when the situation that has arisen on the basis of the perceived state of the DTS turns out to be conscious. And awareness of the situation, as you know, occurs when establishing the sequence of the list of key questions and options for a set of answers. At the same time, it should be emphasized that there is a set of answers and their sequence, indicating a chain of searching for additional information that establishes/confirms the correct directions of judgments or refutes them. Thus, a sufficient level of PTA and, as a result, rational DMP—SFA, formed on the basis of conclusions obtained by argumentation based on the use of a set of judgments as a result of the intellectual activity of the operator. However, intellectual activity, as we know, has two forms: dogmatic and critical.

Dogmatic thinking is usually associated with a ban on discussing the accepted norms of rules, postulates, laws that define permissible situations of the state of DTS. It is necessary in relation to the DMP in cases where these norms, rules, etc., must be strictly observed. Dogmatic thinking presupposes the only form of DMP—SFA, which excludes comparison/consideration of possible options for DMP, independence and freedom of choice. The chain of formation of rational DMP—SFA in such cases is reduced to awareness of the situation corresponding to the regulated rules and extraction from the knowledge base of the corresponding variant of the DMP—SFA. The process of thinking here is reduced to automatism—closing the circuit at the level of "perception—identification—extraction from memory of the corresponding decision". The richer the set of skills indicating the presence of options for such automatism, the more complete is the knowledge base of this sth operator. However, even in this variant, the variant of dogmatic thinking, the key point is awareness of the situation, its identification. And this is almost always associated with the concept of thinking in a broad sense. Thinking represents a stream of concepts, phenomena not directly given. Such a flow of thinking in itself cannot but be clear, maintain consistency and controllability. The professionalism of the operator here lies in the use of reflection—self-control, which allows you to organize and systematize the course of thoughts and judgments. Thus, reflection, as "observation of the mind over its activity" is also characteristic of the dogmatic, based on acute compliance with rules and regulations, variant of thinking. Therefore, it can be argued that reflection, as a phenomenon of self-control, is a necessary mandatory element of all types of thinking, especially important in the preparation and implementation of DMP—SFA, and in all intellectual activities of ATS operators.

It is self-control—reflection that contributes to the awareness of the situation, a clear and consistent construction of a number of key questions, the purposefulness of searching for the necessary information—answers, argumentation of decisions made or denial of less suitable options. It is also important to notice the emotional coloring of such a process that appears at the same time, which consists in acquiring a sense of confidence.

Reflexive thinking plays a special role in the variant of critical thinking. In the process of making decisions, it allows you to acquire the necessary sense of confidence, while forcing the operator to consider the reasons or reasons for his confidence and its logical consequences. This already means reflexive thinking—thinking in the best and brightest sense. The ability of self-control in the course of reasoning using critical consideration of them just forms a critical version of thinking.

Critical thinking acquires a special role in the case of analysis of the management process in such complex active systems as ATS. The fact is that the presence of free will among the operators of such systems leads to the need to consider a special type of organizational management of them—reflexive management. The conflict situations arising here, when the interests of a number of operators in the ATS structure do not coincide or are opposite, require the management groups of operators, for example, the flight manager in the working shift of air traffic control, to develop solutions based on reflexive management. The theory of reflexive control gave rise to one of the variants of mathematical game theory—reflexive games [8, 9]. The results of such a theory, along with the theory of games with non-opposing interests, cooperative games and the theory of compromises, of course, should be used in modeling the activities of both the leaders of the ATS operator groups and the activities of the operators themselves. However, here we are talking about cases of DMP—SFA carried out "manually"—without the use of computer systems, but using perhaps preliminary developments, so-called home analysis or developments obtained during simulator studies and in the process of professional training. A special role here is usually played by rapid reflection, which is often found in practice, when the DMP—SFA is carried out on the basis of the subjective perception of the situation by the manager in the state of DTS, taking into account the individual characteristics of the operators, which he also perceives subjectively.

It is clear that the DMP—SFA in this case will be the closer to rational, the more critical thinking will be used in such a quick reflection. Any kind of cognition, any field of knowledge, like any science, operates with such a generalizing phenomenon as the concept by which the representation of the phenomenon under study is carried out. It is also characteristic of critical thinking. Concepts are formed in the process of professional training and in the activity of operators by such a phenomenon as grasping and, since short-term memory is limited, the process of generalizations in the activity of operators leads to the formation of certain images, for example, which in engineering psychology are called "chunks" [3]. In the DMP activity, the operator operates with "chunks" that have their own descriptions, terms, definitions, both in natural and specially professional, for example, in the language of radio exchange [3.15]. The use of "chunks" allows you to perform speed in the construction of judgments—the foundations of DMP, which are the basis and the most important tool of critical thinking. It is in judgments that insignificant details are separated from significant ones, forming the essence of phenomena, holistically, in synthesis representing the available information or testifying to its lack. Critical thinking, as the main tool of DMP, either presents evidence confirming judgments, or indicates the need for additional information, allowing you to formulate key questions. This is the phenomenon of "judgment delay", the essence of which consists in awareness, synthesis of available information, sorting through possible DMP—SFA and assessing their consequences.

Mastering the skills of "delaying judgments", creating the necessary set of "chunks", the ability of critical thinking and reflection in the analysis of judgments—all these are the most important elements of the PTA and rational DMP—SFA of the

ATS operator, bringing them to automatism, indicating that they have achieved high professional quality.

8.3 Modeling of the Procedure for Assessing the Volitional Tendencies of ATS Operators

In the functioning of active transport systems (ATS), it is possible to distinguish such tasks as the organization of the activities of a separate vehicle traffic control authority, the organization and regulation of vehicle flows (carried out by several traffic control authorities), logistics management of transport processes at various levels, as well as the task of professional training of ATS operators. These tasks have a common feature—several (two or more) participate in the relevant decision-making processes active elements. Appropriate management and functioning mechanisms can be built on the basis of cooperative game models [8, 9].

To build models of ATS functioning, it is necessary to have a description of the utility functions (preferences) of its active elements. Utility theory approaches make it possible to take into account the "economic" component of the concept of utility. However, the activity of the ATS operator, as well as the educational activity of active elements in professional training, apparently cannot be described only within the framework of their economic behavior.

Tasks related to real activities during training, as well as management of both flows and the movement of individual vehicles are tasks that require an appeal to the volitional regulation of active elements—trainees, ATS operators. Such tasks can be represented as the generation of an action that is subjectively necessary and consciously accepted for execution, but not motivationally secured (actions with a lack of motivation to begin). The reason for such a deficit may be, firstly, the polymotivated nature of real human activity, and secondly, the presence of conflict between individual motives. The purposeful nature of the activity in this case is formed on the basis of the principle of dominance. Then, when constructing the preference function of the active elements of the TSy, it is also necessary to take into account their motivational and volitional tendencies.

The preference function determines the meaning-volitional processes of choosing the dominant motivational tendency and the necessary speech-functional acts. The following two questions arise: how to quantify volitional selection processes, and how to build a preference function of active vehicle elements based on these estimates.

There are a number of methods for assessing the state of volitional regulation of active elements, as well as options for their complex application [4], such approaches allow determining the degree of severity of dominant (strategic) motivation. However, if the goal corresponding to strategic motivation lies in relation to the starting point of planning behavioral acts in the area of still objectively intangible achievability (there are no clear criteria and formally defined performance indicators), then it is also necessary to have a vector of strategic dominant motivation (SDM). The processes of volitional regulation corresponding to this vector will ensure the growth of the energy potential of strategic motivation and stimulate behavioral acts to its satisfaction. Volitional tendencies that do not correspond to the SDM vector not only do not ensure the growth of the energy potential of strategic motivation, but can also lead to the emergence of a negative energy potential—loss of awareness of subjective necessity.

During the period of initial vocational training or professional adaptation, the active element is still not clearly able to determine the vector of SDM. It can be assumed that experienced specialists (experts, instructors) have the clearest ideas about SDM. Then, when assessing the level and effectiveness of the semantic-volitional processes of active elements, it is necessary to take into account the degree of their compliance with the SDM vector based on the opinions of experienced specialists.

To construct such a procedure, one can use the method of multi-criteria selection of alternatives based on the compositional rule of aggregation of alternative descriptions with information about the preferences of experts given in the form of fuzzy judgments [10]. In this case, a fuzzy assessment of the degree of consistency of volitional tendencies of the sth active element can be carried out as $F\left(\tilde{G}_s\right) = \frac{1}{\alpha_{\max}} \int_0^{\alpha_{\max}} M\left(G_s^{\alpha}\right) d\alpha$, defined on the basis of a compositional inference rule in a fuzzy environment: $\mu_{\tilde{G}_s}(e) = \max_{\omega \in W}\left(\min\left(\mu_{\tilde{G}_s}(\omega), \mu_{\tilde{D}}(\omega, e)\right)\right)$.

Here $\mu_{\tilde{D}}(\omega, e) = \min_{\omega \in W}\left(\mu_{\tilde{H}_s}(\omega, e)\right)$—general functional solution; $\mu_{\tilde{H}_s}(\omega, e) = \min_{\omega \in W}\left(1, \left(1 - \mu_{\tilde{A}_j}(\omega) + \mu_{\tilde{B}_j}(e)\right)\right)$—implication of fuzzy sets $\tilde{A}_j \subset U_1 \times U_2 \times \cdots \times U_N = W$ и $\tilde{B}_j \subset E = [0, 1]$, corresponding to the statements of experts f_j : $X = \tilde{A}_j \Rightarrow S = \tilde{B}_j$; $\tilde{C}_s$—fuzzy estimates of volitional tendencies of the sth active element by N indicators characterized by linguistic variables $X_i\left(i = \overline{1, N}\right)$, taking values $\tilde{A}_{ij}$ in the jth judgment of experts (information fragment).

8.4 Modeling of Motivation Dynamics and Formation of PTA of ATS Operators

The professional thinking ability (PTA) of specialists of active transport systems (ATS) refers to the key fragments of activity (KFA), because it includes such basic elements of activity as the ability to identify significant informative signs of changes in the dynamic transport situation and critical changes in the condition of the vehicle, as well as to extract information from memory corresponding to these signs and carry out the process decision-making [5, 6, 11, 12]. A special role in the formation of PTA is played by such a motivating force as the initial motivational excitement of the student at the beginning of his initial professional training [3.15]. Mathematical modeling of such motivational arousal and its dynamics will allow us to judge the behavior of the trainee for the position of ATS operator, which is determined by the current value of motivational arousal.

The current value of motivational arousal is characterized by the vector value of the dominant motivation. The mediated value of the dominant motivation can be the correctness (degree of optimality) of the decision-making process carried out by the learner as a result of the assessment of his PTA.

Purposeful formation of PTA in the process of initial professional training of ATS specialists significantly reduces the period of formation of trainees as specialists with high professional qualities [5, 6, 13]. In order to increase the effectiveness of training, it is important not only the formal use of modern teaching methods and tools, but also to take into account and use the determining role and influence of dominant motivation, its dynamics in the professional training of ATS specialists [2, 5].

A special role here is played by the positive dynamics of the dominant motivation, the source of which is the main need of the student—the achievement of high professional quality (PQ). An attempt to study such a problem as the construction of a mathematical model of the dynamics of the dominant motivation of the student in the process of initial vocational training, taking into account its influence on the formation of PTA, is devoted to this section.

Remaining within the framework of the functional systems of P.K. Anokhin-K.V. Sudakov, it can be argued that the dominant motivation is the determining factor of the integral dynamic state of the student, the state of his readiness for certain actions in each of the training situations S_l $\left(l = \overline{1, L}\right)$, generated by simulators or simulators by creating a DTS. Such actions are called speech-functional acts. The mathematical model of motivation that determines the behavior of the sth trainee, then, is a model of a vector force field defined on the space of situations and a set of subjective alternatives to the $\mathrm{SFA}^{(s)}(S_l)$.

Thus, there is a vector dominant measurable motivation that determines, along with the key information about DTS $P^{(s)}(F_k)$, information extracted from memory $\Phi^{(s)}(F_k)$ and emotional relationships $R^{(s)}(F_k)$, behavior—the choice of the s-m learner from a variety of alternatives to his preferred SFA $^{(s)}$ in the situation S_l (Fig. 8.3). Formation of the current value of the motivational field, characterized by strength $F^{(s)}(t_k)$ motivational arousal at the moment of time t_k, allows us to judge the dynamics of the subsequent formation of the dominant motivation—already in the learning process, taking into account the mutual influence of the main components of PTA (blocks 3, 4, 6, 7, 8). It is also appropriate to talk about the influence of the environment on this process—the environment in which the sth trainee is located. It is clear that in the model (Fig. 8.3) the influence of the teacher–instructor is explicated. However, it is not only it that affects education—the dynamics of the motivation of the sth trainee. His behavior and the type of utility functions are further determined by the intentions caused by the desire to "not deceive the hopes" of his environment, as well as by the influence of the generally accepted concept of PQ in this environment and his own ideas about the parameters of such PQ. All these influences are carried out through blocks of perception and blocks of emotional attitude (block 5), memory (block 3‴), in which the base of goals is allocated—fragments of PQ (block 3′), the base of already mastered behavior programs—$\Sigma(\Phi) \to \Phi(J_\Sigma, I_{ij})$ (block

3″), presented in Fig. 8.4 in the form of a functional diagram of the mechanisms of self-learning of the *s*th trainee and the formation of PTA under external influences of the environment and the teacher–instructor.

So, an idea of the dynamics of the power of motivational arousal in a situation emerges S_l $\left(l = \overline{1, L}\right)$ under the influence of the following three most important factors: the influence of the environment—$f_1^{(s)}(t, F, S) = \frac{dF_{окр}^{(s)}\left(t_k, F_k^{(s)}, S_{l,k}\right)}{dt}$, the impact of one's own ideas about PQ fragments—$f_2^{(s)}(t, F, S) = \frac{dF_{соб}^{(s)}\left(t_k, F_k^{(s)}, S_{l,k}\right)}{dt}$, change under the influence of a teacher–instructor—$f_3^{(s)}(t, F, S) = \frac{dF_{инс}^{(s)}\left(t_k, F_k^{(s)}, S_{l,k}\right)}{dt}$.

Thus, a mathematical model of motivation dynamics can be represented as:

$$\frac{dF^{(s)}(t_k)}{dt} = \sum_{i=1}^{3} f_i^{(s)}(t_k, F_k^{(s)}, S_{l,k}), \text{ by } k = \overline{1, N}, t_1 = 0, \ t_N = T. \tag{8.4}$$

The mathematical model of motivation dynamics in the form of a differential equation allows, when solving it, to conduct in-depth studies of the influence of various environmental parameters of educational and educational influences, the *s*th trainee's own characteristics on changes in his motivation, and hence his behavior in the process of initial vocational training. This is possible with the formalization of each of the terms of the right side of Eq. (8.4).

At the same time, it is necessary to take into account some cyclicity of the decision-making processes of the *s-m* trainee under the indirect influence of the dominant motivation factor (Fig. 8.4), determined by his actions that determine the values $F_{k-1}^{(s)}$ and $F_T^{(s)}$ [5]. A cycle appears $t_1 \leq t_k \leq T$ with the final (measured) value $F^{(s)}(T)$.

A difficult task arises: according to the observed actions of the *s*th trainee ($\mathrm{SFA}^{(s)}(t_k, F_\Sigma, S_{l,k})$), to find the best current educational and educational influences determined during the professional selection and its previous characteristics in conditions of uncertain environmental parameters, ensuring $\frac{dF_k^{(s)}}{dt} > 0$.

A solution to such a problem (rather, a problem) has not yet been found. Indeed, it is so large scale and its search has an extensive history, giving rise to a whole list of sciences and scientific directions—from pedagogy and psychology to optimization of process management in hierarchical active anthropocentric dynamic systems. Such a solution would allow us to give an answer to the problem posed here about changing the motivation of the *s*th trainee [taking into account all the influences f_i, $\left(i = \overline{1, 3}\right)$] to obtain a known increment of knowledge, skills and abilities at a known cost of labor, money, time, etc. In other words, to get an answer to the question about the dynamics of the motivation process and the transaction costs of vocational training with the formation of the necessary level of PTA.

The analysis of differential equations (8.4) should be compared with available experimental studies that clearly demonstrate the presence of at least two phenomena:

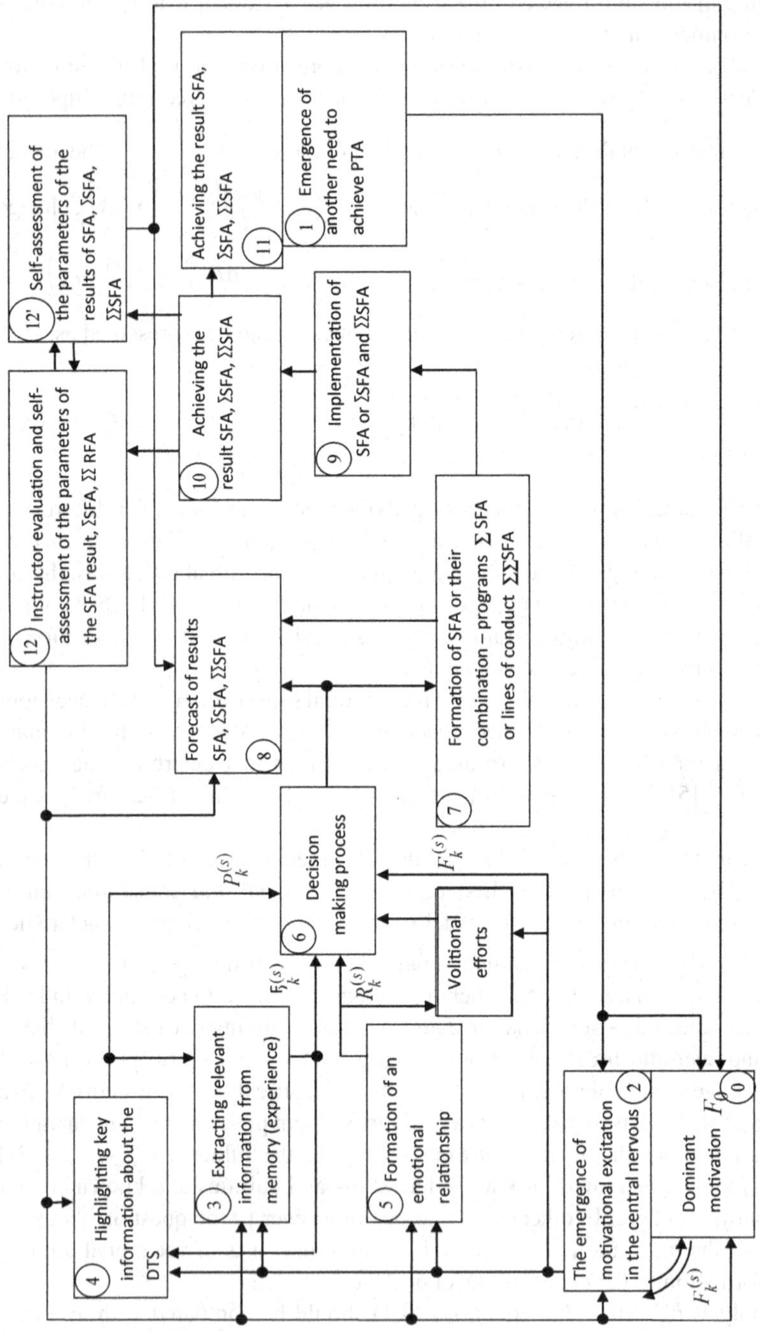

Fig. 8.3 Functional diagram of the mechanisms of formation of motivation in the *s*th trainee, the main components of PTA and SFA

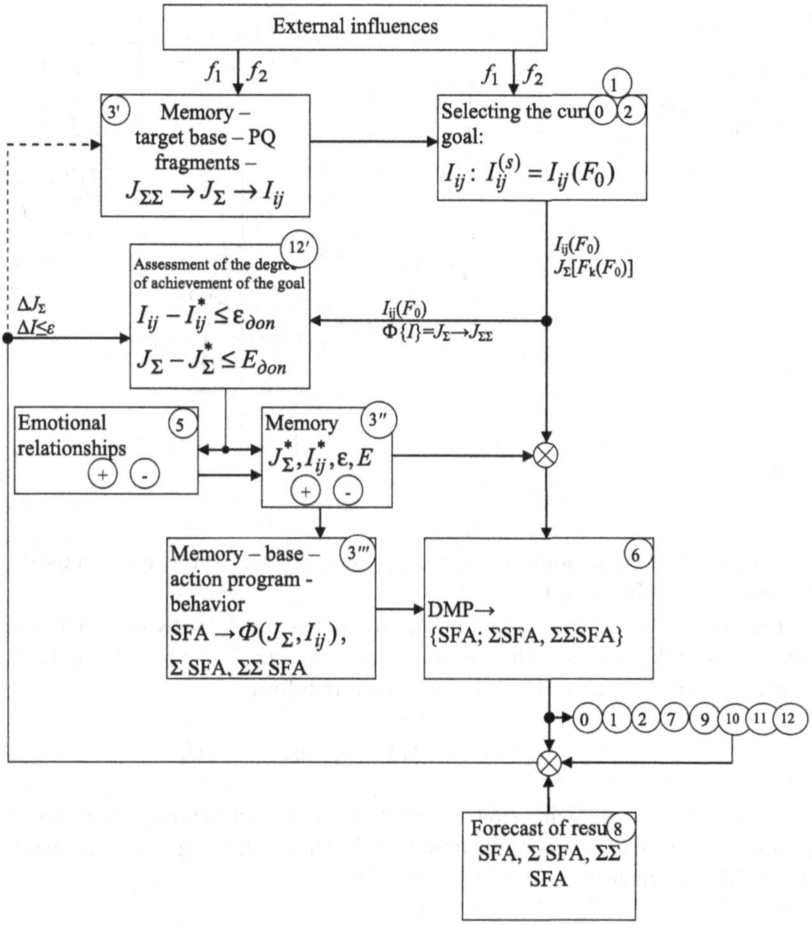

Actions – SFA – current goal - $I_{ij}^{(s)}(F_0^{(s)})$ – motivation at the k-th step - $F_k^{(s)}$.

Action program – ΣSFA – program - $J_{\Sigma}^{(s)}[F_k^{(s)}(F_0^{(s)})]$.

Behaviour – ΣΣSFA – PQ - $J_{\Sigma\Sigma}^{(s)}[F_k^{(s)}(F_0^{(s)})]$.

Memory - (3') (3'') (3''')

PQ – high professional qualities

Fig. 8.4 Functional diagram of the mechanisms of self-learning of the *s*th trainee and the formation of PTA under the influence of motivation under external influences of the environment and the teacher–instructor

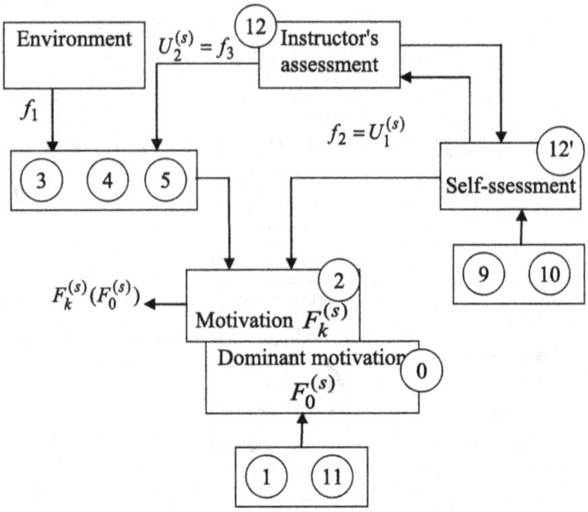

Fig. 8.5 Presence of two control actions and disturbances in the formation of motivation dynamics of the *s*th trainee

the nonlinearity of equations (8.4) and the presence of at least two types of oscillatory processes in the solution (Fig. 8.5).

The first of them with a relatively high frequency is caused by the impact of self-assessment of the results of the *s-m* trainee and is carried out under the influence of "management" in the "control loop" of motivation formation:

$$U_{1k}^{(s)} = f_2(F_k^{(s)}, t_k, S_k) = b_1^{(s)}(t_k, F_k^{(s)}) \cdot U_1^{(s)} \tag{8.5}$$

The second circuit—with a lower frequency—causes a change—the management of motivation formation under the influence of indirect "management" on the part of the teacher–instructor:

$$U_{2k}^{(s)} = f_3(F_k^{(s)}, t_k, S_k) = b_2^{(s)}(t_k, F_k^{(s)}) \cdot U_2^{(s)}(t_u). \tag{8.6}$$

Taking into account these experimental facts, it can be reasonably assumed that the psychophysiological nature of the processes of motivation dynamics of the *s*th trainee can be modeled by nonlinear non-stationary control systems with processes of different rates (Fig. 8.5).

Then the change in the motivation of the *s*th trainee is described by a nonlinear non-stationary equation:

$$\frac{dF_k^{(s)}}{dt} = f_1^{(s)}(t_k, F_k^{(s)}, S_{l,k}) + b_1^{(s)}(t_k, F_k^{(s)}) \cdot U_1^{(s)} + b_2^{(s)}(t_k, F_k^{(s)}) \cdot U_2^{(s)}(t_u), \tag{8.7}$$

where $F_{k=1,N}^{(s)}$—output vector variable available for measurement at $k = 1, N$; $F_k^{(s)} = F_1^{(s)}$—initial conditions; $U_1^{(s)}$ и $U_2^{(s)}$—control actions that vary within $\Omega_{1,2}$—"management resource", Regarding functions f_1, b_1, b_2 it is assumed that they are continuous by t_k, $F_k^{(s)}$, $S_{l,k}$ and limited by limits 0 and $\Omega^{(s)}_{\max(f_1,b_1,b_2)}$; $t_u = \Theta \cdot t$, where t_u—the pace of the teacher–instructor's impact, t—the rate of transients under the influence of only $U_{1k}^{(s)} = f_2 = b_1 U_1^{(s)}$, $\Theta \geq 100$—the degree of separation of the rates of transients in the subsystem of fast oscillatory movements (SFOM) and in the subsystem of slow oscillatory movements (SSOM).

Thus, already at this stage of research, it can be concluded that the most influential influence on the change in the motivation of the trainee will be such an impact of the teacher–instructor, which falls into the resonance of "own" oscillatory movements in the SFOM, determined by the personal characteristics of the sth trainee.

Output vector variable $F_k^{(s)}$ it is formed by a number of components—types of motivation (i.e. $F_k^{(s)} = (X_1^{(s)}, \ldots, X_p^{(s)})$), taking into account both the material (various kinds of remuneration) and the non-material component. For our case, the most important component of the motivation vector is the increment of knowledge, skills and abilities that determine the level of achieved PTA, and are necessary to achieve the main need—PQ. The payment for such an increment is the cost of training in the form of a vector formed by cost, time, labor, etc., costs, forming what is called transaction costs in the economic literature $3^{(s)}(t) = (y_1^{(s)}, \ldots, y_m^{(s)})$.

At the moment t_k there is a situation in which the sth trainee has defined $F_k^{(s)}$ and $3^{(s)}(t)$, forming pairs at every moment of time t $[F_k^{(s)}; 3^{(s)}(t_k)] \in \mathfrak{Z}^{(s)}_{\max}$, on the set of which it is possible to represent the utility function for the sth trainee:

$$\Pi^{(s)}[F_k^{(s)}; 3^{(s)}(t)]. \tag{8.8}$$

Let this function be such that its large values correspond to large values of the learner's preferences, and they are continuous up to the second derivative.

Most often in the literature on economics and management, there are types of utility functions represented by the following dependencies:

$$\left.\begin{aligned} \Pi_1^{(s)} &= \varphi^{(s)}(X) + a^{(s)}[T - \psi^{(s)}(y)]; \\ \Pi_2^{(s)} &= \varphi^{(s)}(X) + a^{(s)}[T - \psi^{(s)}(y)] + b^{(s)}[T - \psi^{(s)}(y)]^2; \\ \Pi_3^{(s)} &= \ln \varphi^{(s)}(X) + a^{(s)} \ln[T - \psi^{(s)}(y)], \end{aligned}\right\} \tag{8.9}$$

where $a^{(s)} > 0$—ratio, $\varphi^{(s)}(X) \geq 0$—motivation change function, $\psi^{(s)}(y) \geq 0$—cost function.

It is safe to assume that the emergence of motivation, as an increment of knowledge, skills and abilities, requiring appropriate costs, is associated with the utility function:

$$F_k^{(s)}(F_0, \Phi_k, R_k, P_k) = F^{(s)}(x, y) = \frac{dx}{dy} \approx \frac{\Delta F_k^{(s)}}{\Delta 3^{(s)}(t_k)}, \tag{8.10}$$

that is, the value of the motivation function $F_k^{(s)}(x, y)$ at some point with the values of the increment of knowledge that increases the level of PTA $x(t_k)$, and costs $y(t_k)$

is equal to the tangent of the slope angle of the tangent curve of the utility function at this point: $F_k^{(s)}(t_k) = \frac{\mathrm{d}x(t_k)}{\mathrm{d}y(t_k)}$.

So, for example, for the utility function $\Pi_3^{(s)}$ from (8.10) should:

$$F^{(s)}(t_k) = \frac{a^{(s)} \cdot \varphi^{(s)}[x(t_k)]}{T - \psi^{(s)}[y(t_k)]}. \tag{8.11}$$

The actions of the sth trainee can be considered rational if they are defined by a vector function of the demand for knowledge:

$$Z^{(s)}(c,x) = \underset{(cx)\le 3_{\max}^{(s)},\ c,x \in \mathfrak{Z}_c^{(s)},\mathfrak{Z}_x^{(s)}}{\arg\ \max}\ \Pi^{(s)}(x), \tag{8.12}$$

where $c = (c_1, c_2, \ldots, c_n)$—prices for various types of acquisition-increments of knowledge, skills, skills; $3_{\max}^{(s)}$—the marginal value of the cost of incrementing knowledge. Now it becomes clear how any additional costs will be used to increase knowledge, i.e. increase the level of PTA, since motivation will also be determined by dependence:

$$F_k^{(s)} = \frac{\mathrm{d}\Pi^{(s)}(x)}{\mathrm{d}c}, \tag{8.13}$$

which in economics is called “direct elasticity”.

When the value $\Pi^{(s)} = \Pi_3^{(s)}$and get the value $F^{(s)}(t_k)$ from (8.10), the vector field (8.13) can also be interpreted as economic motivation, and the condition (8.12) as a condition for rational transaction costs. For a more in-depth economic study, it is important to trace the change in the demand function for the process of knowledge increment (8.11) depending on the change in the values of marginal costs $3_{\max}^{(s)}$, as which it is interesting to consider state budget financing, as well as depending on changes in the price vector $c = (c_1, c_2, \ldots, c_n)$.

It is important to note that at some initial moment for the sth trainee $t_k = t_k^{(0)}$, knowing the price vector c and the marginal cost values for it $3_{\max}^{(s)}(t_k^{(0)})$, from (8.13), based on (8.12), it is possible to determine the initial conditions for the vector differential equation of motivation dynamics (8.4) and (8.6), i.e. the vector $F^{(s)}(t_k^{(0)}) = F_r^{(s)}(t_1)$, where $r = \overline{1, R}$—types of motivation for the sth trainee.

As already noted, the dynamics of motivation of the sth trainee, formed taking into account the structure (Figs. 8.3 and 8.4), significantly depends on the influence of the environment, and first of all—people close to him, as well as on the educational and training influences of the teacher–instructor (8.4). At the same time, each of the significantly influencing parties is characterized by the presence of its own motivation, different from the motivation of the sth trainee himself $F^{(s)}$.

Let the situation be in the first approximation S_l $\left(l = \overline{1, L}\right)$ it is stable enough and does not change for a certain period, exceeding at least the rate of transients of the subsystem of slow oscillatory movements by several orders of magnitude (Fig. 8.5). This allows us to consider the forces influencing the dynamics of the motivation of the sth trainee independent of S_l. It can also be assumed that the change in motivation of this type $F^{(s)} = F_r^{(s)}$ (the desire to gain an increment of knowledge, skills, skills) does not depend on changes in his other motivations $\left(r = \overline{1, R}\right)$.Then the vector Eq. (8.7) takes the form of a system of differential equations:

$$\frac{dF_r^{(s)}(t_k)}{dt} = f_1^{(s)}(t_k, F_r^{(1)}, \ldots, F_r^{(i)}, \ldots, F_r^{n}) + f_2^{(s)}(t_k, F_r^{(s)}) + f_3^{(s)}(t_k, F_{rs}^{(И)}, F_r^{(s)}), \tag{8.14}$$

where $F_r^{(i)}$—motivation of the rth type of the ith person from the environment of the sth trainee $\left(i \neq s, i = \overline{1, n}\right)$, determining his attitude to this particular type of motivation of the sth trainee—$F_r^{(s)}$; $s = \overline{1, N}$, N—number of trainees; $F_{rs}^{(И)}$—the motivation of the teacher–instructor in relation to the sth trainee according to his rth aspiration.

If we determine by polling the degree of influence of each of the $i = \overline{1, n}$ persons of the environment of the sth trainee as weight coefficients $\omega_i^{(s)}$, then the function $f_1^{(s)}$ it can be represented in the simplest case by a stationary function of linear interaction in the form:

$$f_1^{(s)} = A_s \left[\sum_{i=1}^{n} \omega_i^{(s)} F_r^{(i)} - \omega_s F_r^{(s)} \right], \tag{8.15}$$

where $\omega_i^{(s)} \geq 0$; $\sum_{i=1}^{n} \omega_i^{(s)} = 1$; A_s—the coefficient of influence of the environment and $A_s > 0$; $\omega_s >> \omega_i^{(s)}$, but $\omega_s < \sum_{i=1}^{n} \omega_i = 1$.

One's own ideas about the fragments of the PQ at this period of the sth trainee can be represented by the presence of the "ideal" rth type of motivation for him $F_r^{*(s)}$. Then

$$f_2^{(s)} = B_s [F_r^{*(s)} - F_r^{(s)}], \tag{8.16}$$

where $F_r^{(s)}$—the current (desired) value of the motivation of the sth trainee; B_s—the coefficient of individuality, and $B_s \geq 0$, $F_r^{*(s)} \geq 0$.

Аналогично для силы воздействия преподавателя-инструктора на s-го обучаемого в самом простом варианте может быть использованы функция вида:

$$f_3^{(s)} = D_s [F_{rs}^{(И)} - F_r^{(s)}], \tag{8.17}$$

где $D_s \geq 0$—коэффициент обучения, и $F_{rs}^{(И)} \geq 0$—мотивация преподавателя-инструктора в отношении s-го обучаемого.

Stationary solution $F_r^{(s)} = F_r^{*(s)} \approx F_r^*$ systems (8.14) with educational and psychological equilibrium, determined from the condition $\frac{dF_r^{(s)}}{dt} = 0$. Then from (8.14) with values of f_i $(i = 1, 2, 3)$,equal to those given in formulas (8.15)–(8.17), a system of linear algebraic equations is obtained:

$$(E_s + \omega_s)F_r^{(s)} - \sum_{i=1}^{n} \omega_i^{(s)} F_{rs}^{(i)} = E_s K_s, \tag{8.18}$$

where

$$\left.\begin{aligned} E_s &= \frac{1}{A_s}(B_s + D_s); \\ K_s &= \frac{1}{E_s}\left(\frac{B_s}{A_s} F_r^{*(s)} + \frac{D_s}{A_s} F_{rs}^{(И)}\right). \end{aligned}\right\} \tag{8.19}$$

Thus, the stationary solution of the system (8.19):

$$F_r^{(s)} = \frac{K_s E_s + \sum_{i=1}^{n} \omega_i^{(s)} F_{rs}^{(i)}}{E_s + \omega_s}, \quad s = \overline{1, N}. \tag{8.20}$$

On condition $B_s >> D_s > A_s$ get $F_r^{(s)} \to F_r^{*(s)}$. On condition $D_s >> B_s > A_s$ get $F_r^{(s)} \to F_{rs}^{(И)}$.

It can be shown that such a stationary solution is stable in the sense of Lyapunov if the conditions are met:

$$\left.\begin{aligned} &\omega_s + \tfrac{1}{A_s}(B_s + D_s) \geq 1; \\ &\sum_{i=1}^{n} \frac{\omega_i^{(s)}}{\omega_s + \frac{1}{A_s}(B_s + D_s)} \leq 1; \\ &\sum_{i=1}^{n} \omega_i^{(s)} \leq \omega_s + \tfrac{1}{A_s}(B_s + D_s). \end{aligned}\right\} \tag{8.21}$$

However, for our task, the case of an increase seems to be more important $F_r^{(s)}$ $\left(\frac{\partial F}{\partial t} > 0\right)$, i.e. $(E_s + \omega_s)F_r^{(s)} - \sum_{i=1}^{n} \omega_i^{(s)} F_r^{(i)} = E_s K_s$, where, as before: $E_s = \frac{1}{A_s}(B_s + D_s)$ and $K_s = \frac{1}{E_s}\left(\frac{B_s}{A_s} F_r^{*(s)} + \frac{D_s}{A_s} F_{rs}^{(И)}\right)$.

Under the conditions $B_s >> D_s > A_s$ we get instead (8.20):

$$F_r^{(s)} > \frac{K_s E_s + \sum_{i=1}^{n} \omega_i^{(s)} F_{rs}^{(i)}}{E_s + \omega_s} \text{ and then } F_r^{(s)} > F_r^{*(s)},$$

that is, the desire for PQ for the fanatically aspiring sth trainee leads to some "overlap" in the growth of the rth type of motivation and to uncontrolled (or poorly controlled) on the part of the teacher–instructor of its growth.

Here the relations between the coefficients of the influence of the environment come into force A_s, personalities B_s and training D_s. With their help, as well as by interviewing and selecting the values of the weight coefficients $\omega_i^{(s)} \geq 0\ \left(i = \overline{1, n}\right)$ the influence of persons on the sth trainee from his environment, and the values of $\omega_s > 0$, you can simulate any real situation $S_l\ \left(l = \overline{1, L}\right)$, because all the named coefficients are the same in each of them.

If, for example, we consider the motivation of a teacher–instructor to be a determined dependence

$$F_{rs}^{(И)} = \frac{d_И^{(s)} \cdot \varphi_И^{(s)}[x(t_k)]}{T_И - \psi_И^{(s)}[y(t_k)]}, \tag{8.22}$$

where, as before, $d_И^{(s)} = D_s$ and $T_И$—the corresponding parameters of the utility function (8.9), what it should be (or become) according to the instructor, then:

$$\Pi_3^{(s)(И)} = \ln \varphi_И^{(s)}(X) + d_И^{(s)} \ln[T_И - \psi_И^{(s)}(y)]. \tag{8.23}$$

Thus, the stationary solution of the system (8.20) characterizes the final value of the increment of motivation as a result of taking into account all influences, which, if the conditions (8.21) are met, is stable for this situation $S_l\left(l = \overline{1, L}\right)$. However, economic motivations of the form (8.13), when they characterize the utility function of the sth trainee, significantly depend on the situation and are related to the utility function $\Pi_3^{(s)(И)}$(8.23).

This leads to the conclusion that the mathematical model of motivation dynamics should be more complex and, in principle, should depend on changes in the situation, for example, taking into account costs in the form of functions $\psi(y)$ and the increment of motivation $\varphi(x)$.

For a more complete representation of the dynamics of motivation, it is necessary to take into account experimental data on the values of the parameters of Eqs. (8.8) and (8.14), for example, at least those that generate a variety of processes and characterize the transients of the "internal" part in the motivation formation system—blocks 0, 2, 12', and "external"—blocks 12, 3, 4, 5, 2, 0 in Fig. 8.5.

It has been experimentally established that if the tempo of the "inner" part is determined by a period of seconds, then the tempo of the "outer" is characterized by a time period of tens of minutes. This allows us to consider the analysis of the solution of the nonlinear non-stationary differential equation (8.8) on separate clock cycles—periods—after each "external" influence of the teacher–instructor, the consequence of which will be a new initial condition for the equation describing the dynamics of motivation of the sth trainee.

So far, it can be stated that the available experimental material is still insufficient to guarantee the full adequacy of solutions of equations of the form (8.8), (8.14) to the processes of motivation dynamics of trainees.

It is possible that the partial differential equations of parabolic and biharmonic type characterizing the nature of the motivation dynamics of the sth trainee will be more adequate.

At the same time, dependencies (8.20) and (8.23) allow us to trace the role of personal qualities and assess the degree of the necessary influence of the teacher–instructor and the environment, if necessary, to change the dynamics of motivation of a particular type in this sth trainee. The result in this case will be the achievement of exactly the required level of PQ on the basis of the development of skills, abilities and knowledge that form (form) the necessary PTA for this sth trainee.

A significant place in the activities of an HAS specialist is occupied by processes related to the solution of mental tasks. As is known, the dominant motivation plays a leading role in programming the activity of any individual, including in programming his mental activity [1, 5, 13]. If it is possible to form a stable dominant motivation in the student to achieve professionally high qualities (PQ), then in the process of professional training, as a result of repeated satisfaction of the same type of local needs that make up the global need to achieve PQ, a common intention arises—the desire to participate, which forms the mechanism of imprinting—memory, skills, abilities and knowledge. In the systemic mechanism of imprinting and strengthening knowledge, one of the most crucial moments of the learning process is the process of extracting, under the influence of the dominant motivation, the experience accumulated in the preparation process and, in particular, the experience, the formalization of the goal and thinking ability. It is this process that determines the qualitative transition from passive learning to active learning—with the expectation of a result—in extracting the future result of professional training at this stage, this exercise, this task. This is how the creative cognitive-activity sphere of a specialist is formed, formed in the form of concepts, skills, skills, judgments, representations, images, etc. The process of mental learning is based on a constant system-quantum action of the acceptor of the results of the student's action, by creating special "images", "tensors", "matrices", "vectors", blocks, etc., stereotypes of knowledge, which are subsequently relatively easy to extract under the influence of dominant motivation. It is as a result of the influence of the acceptor of the result of an action based on feedback that skills are formed according to the necessary speech-functional acts (SFAs) of ATS specialists [2, 5]. At the same time, in the process of professional training, the phenomena of the development of the acceptor of the result of an action are manifested in the form of formation and intuitive activity. The mental activity of a ATS specialist thus consists of the interaction of system–organized components, including blocks of feelings (emotions), motives, memory (goal-setting, programs, representations, fantasies, imagination, etc.), decision-making, will, etc.—with the leading role of dominant motivation, these components and their interrelationships can be considered as the functional basis of processes the thinking of ATS specialists and/or the basics of their PTA. When trying to model the functioning of such blocks, it is necessary to take into account their relationship and the presence of delays in the formation of PTA and dominant motivation. These two interrelated processes are described by different differential equations forming a system. Thinking has received a rather vague definition in a number of philosophical psychological and other publications, although the process of thinking does not yet have an exhaustive scientific interpretation, let alone modeling. It is reliably known that the process of thinking consists of—includes the processes of cognition of data, generalization of them in the form of concepts, images, etc., a system of conclusions, conclusions, methods of solving problems, as well as forecasting. The result of mental activity is the formation of thoughts realized in the form of goal-setting, program moments preceding decision-making processes, where elements of will, emotions and prediction of consequences are already included. Thus, the formed thoughts determine the subjective reflection by the learner of his objectively existing needs that determined the dominant motivation, synthesis and prediction of programs for their satisfaction in interactions with memory through the implementation of SFA. At the same time, it is important to emphasize that the process of thinking, like any process of activity, is carried out by systemquants—from the beginning of the task to its solution, passing through the execution of a certain chain of mental actions.

At the same time, such a chain includes various elements in the case of the process of solving a mental task by a trainee from a type familiar to him from training and in another, where he is faced with a new task. It is in the latter case that the whole chain is turned on, starting with the dominant motivation and, as a rule, with significant efforts, a new functional system is formed, and, under the influence of emotions, it is fixed in memory. The thought process occurs most often with the use of verbal images, most often with internal speech. From the standpoint of the general theory of functional systems P. K. Anokhin and K. V. Sudakov's mental act is accompanied by volitional efforts, which are required especially acutely in cases of uncertainty and the presence of time deficit conditions. These efforts are determined by the processes of afferent synthesis, when external circumstances—the external environment, DTS, etc., come into conflict with the dominant motivation, when memory mechanisms containing moral, social, moral principles—prohibitions that form the area of permissible SFAs are activated. It is clear that a special role in the formation of such an area, as well as the process of mental activity itself, is played by the condition of instructions, instructions and rules necessary for a ATS specialist to comply with. The whole complex of knowledge, skills and abilities acquired by a trained ATS specialist, taking into account the assimilation of instructions, instructions and rules, forms a certain hypersystem of knowledge consisting of a set of functional systems that form them in the process of preparation. On their basis, certain stamps and stereotypes of his mental activity and his knowledge, characteristic only for this ith trainee, are formed. They are, as a rule, very inertial and their changes are difficult to change.

The functional systems formed in the process of professional training, as well as in practice, allow solving the main part of the mental tasks that arise before the ATS specialists. However, in special cases of DTS, as well as in case of failures of transport equipment and/or special environmental phenomena, there are moments that require solving mental problems, for which the functional systems in the central nervous system (CNS) of a specialist are not formed. This happens most often when there were no reinforcing results of the action in the acceptors due to the lack of visual, sensory, signal information—when conditions of uncertainty and time deficit, so characteristic of DTS, arise. And in general, in the activity of a specialist, such variants of events emphasize the importance of simulator and practical training of ATS specialists—on the one hand, and indicate on the other that thinking in practice, relying on the formed hypersystem of knowledge, still goes beyond it, using intuition, heuristics based on existing knowledge, and especially on the will, forcing you to suppress emotions and follow instructions, rules, instructions and moral principles. All this requires the development of PTA properties in order to replenish the hypersystem of knowledge and personality development.

A special role in the thought process that forms the PTA is played by the moments of the emergence of generalizations that allow grouping existing functional systems and creating, in conditions of uncertainty, the necessary—desired goals—the results of programmed actions of the SFA. This process within the framework of the theory of functional systems can be represented as the formation of self-motivation, when the predicted decisions and results of actions based on intuition and/or heuristics in significant—critical time is not fully defined. Thus, for ATS specialists with PQ, along with a hypersystem of knowledge based on clearly delineated feedbacks created by the acceptor of the results of actions, when they—these results are predetermined—are known, a significant role is played by PTA with self-development in the form of the possibility of self-motivation and the required volitional efforts in the process of thinking DMP and SFA. The hypersystem of knowledge and visual sensory signals about DTS and other information in conditions of uncertainty and time scarcity certainly play a significant role, but the inclusion of a conceptual level, a generalized grouping of functional systems, and especially the ability of self-motivation and the inclusion of will in the process of thinking gives significantly new results, both in the study and in the modeling of such processes (Figs. 8.5 and 8.6).

It is also important to emphasize that the formation of PTA is carried out on the basis of a gradual or abrupt (with generalizations) mental process of forecasting and future results, programmed SFAs, its evaluation and signals about it, always inherent in them, in conditions of uncertainty and time scarcity. Such a forecasting process goes on continuously, as an initially holistic process of thinking, in which no stages, elements that make up; the formation of any subsequent step—stage of mental activity is a continuation of the formation of the previous step—stage, etc. Thus, any steps—stages of the living process of thinking are not separated, disconnected from each other; i.e. they are inseparable, continuously and significantly changing, dynamic, forming and developing. This, taking into account the ability of self-motivation and the inclusion of will, is, in our opinion, the main difference between living knowledge and knowledge in intelligent artificial control systems.

Therefore, it can be argued, like everything else in this world, the process of thinking and the formation of PTA is the unity of continuous and discontinuous—discrete. It is clear that modeling such processes is associated with extreme difficulties.

The level of the formed PTA is most often assessed by the ability of the student to issue ideas, SFAs, etc., differing in originality, novelty, while being within acceptable solutions and, most importantly, corresponding to the emerging (set) task. This ability arises on the basis of the hypersystem of knowledge already inherent in this trainee and his possession of moments of self-motivation and the inclusion of will [2]. As

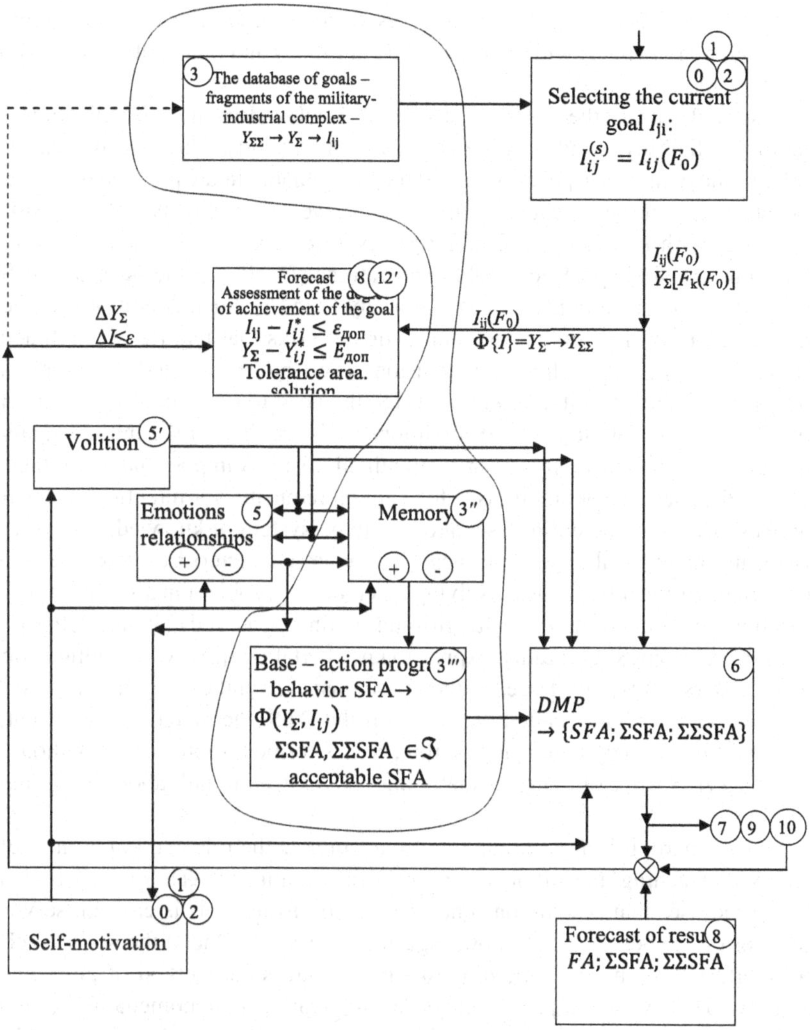

Actions – SFA – current goal -$I_{ij}^{(s)}\left(F_0^{(s)}\right)$ – motivation at the k-th step -$F_k^{(s)}$;

Action program – ΣРФА – program -$Y_{\Sigma}^{(s)}[F_k(F_0)]$;

Поведение – ΣΣSFA – PQ -$Y_{\Sigma\Sigma}^{(s)}\left[F_k^{(s)}\left(F_0^{(s)}\right)\right]$;

Memory is the area of acceptable SFAs (3)(3′)(3″);

Needs – dominant motivation – self-motivation (0)(1)(2)(2′);

Fig. 8.6 Functional diagram of the *s*th self-learning mechanisms DM$^{(s)}$ under the influence of motivation, as well as self-motivation processes in conditions of uncertainty and lack of time

is known, the relationship of such abilities with the general level of intelligence is estimated non–linearly: with low variability at low IQ and high—at high, on average about $r = 0.30$.

A special role in the assessment of PTA is played by the identification of its stability, reliability, validity in conditions of uncertainty—lack of awareness and lack of time, which is especially important and valuable in the properties of PQ ATS specialists. It can be argued that the general level of intelligence IQ is positively associated with PTA and, in turn, a high level of PTA contributes to the ability to determine the finding of acceptable and rational DMP and the issuance of SFAs in conditions of uncertainty and time scarcity [2, 5, 6]. Such conditions put before the learner a set of heuristic and semi-heuristic tasks that require research activity from him, impossible without the inclusion of self-motivation and volitional efforts to "grasp" the holistic situation in the DTS, the need to take into account the entire set of acceptable solutions in the conditions of uncertainty of the initial signals and the absence of forecasts. It is clear that with all this, it is impossible not to take into account the presence of emotions, which are prerequisites for intentions—desires—motives, including the entire arsenal of the hypersystem of knowledge, experience, which ultimately make consciousness. The main characteristics of consciousness here are intentionality, focus on solving an emerging problem in a situation of DTS, variability, continuous fluidity, integrity and continuity, as well as connectedness with flows of emotions, structurality, expression in a certain sequence of emotion components, observability, etc. Here, in the consciousness synthesizing the hypersystem and experience, what is called the image of the DTS, the image of movement, the image of flight of a dynamic air situation, etc., is formed, synthesized. Without such a synthesizing work of consciousness, there is no need to talk about the formation of PTA.

In this regard, it is appropriate to once again note the role and importance of the quality of teaching and the level of simulator training of ATS specialists. It is in these processes that a common fund of semantic formations is created, serving as the basis of a hypersystem of knowledge and experience. One of the ways to achieve the shortest path and the formation of such a fund is the method of joint-dialogic cognitive activity, which just develops self-motivational phenomena in the student. Earlier, the importance of emotional pedagogy was noted, when the knowledge, skills and abilities of the student are sharper, clearer and more reliably implemented in the presence of the emotional influence of the teacher–instructor. It is also important to take into account the role of positive thinking in the process of launching achievement motivation, self-achievement, self-esteem and the effectiveness of the learning process. Optimistically thinking trainees are more successful in the future activities of ATS specialists, who are characterized by conditions of uncertainty, lack of time and which require the presence of the properties of perseverance, perseverance, resourcefulness. An optimistic style of thinking is always associated with a more successful academic performance in learning, a more pronounced ability to self-motivate and include the will in the search for acceptable DMP [2].

Thus, an idea emerges about the role of training and coaching, the importance of the intellect and PTA of the trainee, ways, methods and techniques for their achievement and development. Such conclusions have been made many times before, for example, for specialist operators of such a complex type of ATS as air transport [2, 5, 6], etc. Taking into account the findings, it becomes obvious that there is a need to develop tests that allow assessing the value of the achieved level of PTA and the properties of including self-motivation and will in situations with uncertainty and lack of time in the training of ATS specialists. At the same time, the prognostic validity of such diagnostic tests and their basing on generally encompassing tests of mental development are also important. It is clear that the development, testing and verification of such tests require considerable effort, time and money. The justification for such costs would be the study of transaction costs throughout the professional path of specific ATS specialists who are graduates of a particular educational institution of transport orientation. It is clear that such a battery of tests used in the process of professional training and certification of ATS specialists should in some sense be a continuation and development of the tests used in the professional selection of applicants. A complex of such batteries of tests will guarantee the achievement of a high level of professional qualities by the trainees, the basis of which, as has been shown, is professional thinking ability.

The emerging motivational arousals and the emerging PTA are interrelated, and the cognitive–intellectual component of motivation should prevail over the physical (need-carnal) and include the moral (spiritual) component as one of the main ones in the vector of dominant motivation. Without delving into the process of modeling the problem of choosing between the needs of the spirit and flesh, soul and mind, which form the basis for the formation of dominant motivation, we will take as the basis of the model of motivation dynamics of the r-type differential equation obtained on the basis of (8.14)–(8.17):

$$\frac{\mathrm{d}F_r^{(s)}(t_k)}{\mathrm{d}t} = A_s\left[\sum_{i=1}^{N}\omega_i^{(s)}F_r^{(is)} - \omega_s F_r^{(s)}(t_k)\right] + B_s\left[F_r^{*(s)} - F_r^{(s)}(t_k)\right] + D_s\left[F_{rs}^{(N)} - F_r^{(s)}(t_k)\right] \tag{8.24}$$

where $F_r^{(s)}(t_k)$—the rth type of motivation of the sth trainee at the time (t_k); $F_r^{(is)}$—motivation of the kth type of the ith person relative to the sth trainee: $i \neq s$; $\omega_i^{(s)}$, ω_s—coefficients of influence of the ith persons on the sth trainee and his self-influence; A_s, B_s, D_s—coefficients of the corresponding dimensions characterizing the influence of the environment, individuality, learning ability, respectively; $F_r^{*(s)}$—the self-representation of the sth trainee is the "ideal" degree of his motivation; $F_{rs}^{(N)}$—the motivation of the instructor–teacher relative to the sth trainee according to the rth type of motivation;

The vector differential equation of the dynamics of the PTA of the sth trainee in the PTA component can then be represented by analogy with (8.24) in the form:

$$\frac{\mathrm{d}V_r^{(s)}(t_k)}{\mathrm{d}t} = k_1^{(s)} F_r \Im_{1(s)}\left[A_s, B_s, D_s, \omega_i^{(s)}, \omega_s, F_r^{*(s)}(t_k)\right] \cdot V_r^{(s)}(t_k)$$
$$+ k_2^{(s)} F_r \Im_{2(s)}\left[A_s, B_s, D_s, \omega_i^{(s)}, \omega_s, F_r^{*(s)}(t_k)\right] \cdot I_{\sum r}^{(s)}$$
$$\cdot \left(F_{rs}^N, t_k, \Im_{3(s)}\left[k_3 V_r^{*(s)} - k_4 V_r^{(s)}(t_k)\right]\right) \tag{8.25}$$

where $V_r^{(s)}(t_k)$—the rth component of the volume vector of the hypersystem of knowledge of the sth trainee—his PTA at the time t_k; $k_1^{(s)}$, $k_2^{(s)}$—coefficients corresponding to the dimension characterizing the degree of influence of self-learning and learning when presenting new (for the sth trainee) information $I_{\sum r}^{(s)}$; as is known, [2] the rth component $V_r^{(s)}(t_k)$ physically, it is the inverse of the weighted errors in the rth direction of the knowledge system at the time t_k; $V_r^{*(s)}$—ideal—the desired value of the rth component.

The characteristic type of solution of the system (8.24), (8.25) is shown in Fig. 8.7.

The system of equations (8.24), (8.25) represents different-time changes in the motivation of the sth trainee and the magnitude of his rth component of the knowledge vector caused by both self-motivation, motivation of the environment and the instructor, as well as the influence of the information influence of the instructor–the teacher. Both the type of Eq. (8.25) and its semantic essence are close to the Brooks formula, which describes the process of generating a portion of new knowledge in

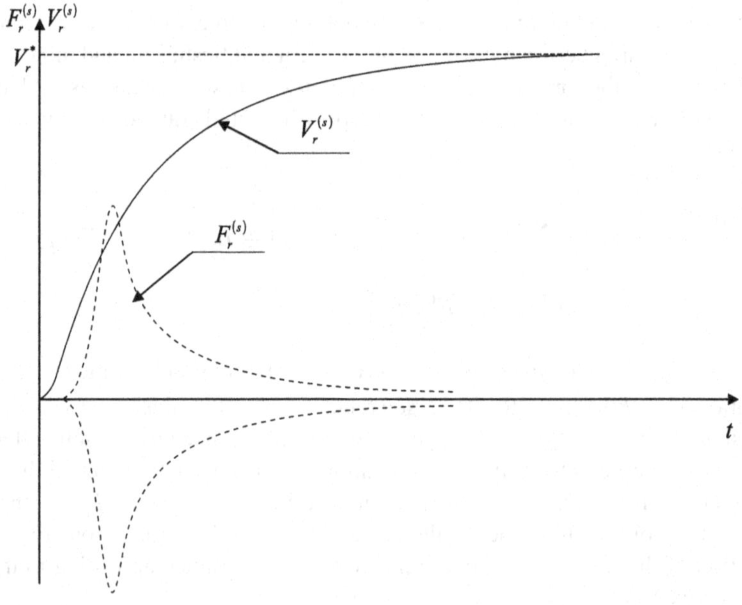

Fig. 8.7 Change of the rth component of the vector of the volume of the knowledge system of the sth trainee in the presence of motivation $F_r^{(s)}(t_k) > 0$

the rth direction based on their previous volume $V_r^{(s)}$ and getting new information $I_{\Sigma}^{(s)} \cdot (D_s, F_{rs}^N, t_k)$ perceived by the sth trainee [2].

A rational management system for such a process of acquiring and generating knowledge can be built taking into account the possibility of parametric invariance of learning outcomes to the "spread"—the uncertainty of the characteristics of the trainees $(s = \overline{1, N})$. The processes of testing knowledge in this case can acquire a dynamic character and is carried out using simulation modeling. If we assume that the probability—the possibility of acquiring—generating knowledge depends on the amount of knowledge already available in terms of training time, with the coefficient $\mu_s = f\left(B_s, D_s, k_1^{(s)}, k_2^{(s)}\right)$, then, according to the law of pure reproduction, such an opportunity is obtained for the sth trainee in the form of a Yule-Farry distribution. Provided that the exponential distribution is used for the study time in the rth direction:

$$P_V^{(s)} = \int_0^{\infty} \mathrm{e}^{-\mu_s t}\left(1 - \mathrm{e}^{-\mu_s t}\right)^{\left(V_r^{(s)} - 1\right)} \cdot \lambda_r^{(s)} \mathrm{e}^{-\lambda_r^{(s)} t} \cdot dt \approx \frac{\lambda_r^{(s)}}{\mu_s} B\left[V_r^{(s)}, \frac{\lambda_r^{(s)}}{\mu_s} + 1\right] \tag{8.26}$$

where μ_s—the coefficient characterizing the individual properties of sth—motivation—learning ability; $\lambda_r^{(s)}$—parameter of the distribution of the training time of the sth trainee in the rth direction.

$V_r^{(s)} = 1, 2, \ldots; B\left[V_r^{(s)}, k_r^{(s)} + 1\right]$—beta function.

$$\text{Denoting } \frac{\lambda_r^{(s)}}{\mu_s} = k_r^{(s)} \tag{8.27}$$

and with $V_r^{(s)} \to \infty$ using Stirling's formula, you can get:

$$B\left[V_r^{(s)}, k_r^{(s)} + 1\right] \to \frac{1}{V_r^{(s)\left(k_r^{(s)}+1\right)}} \tag{8.28}$$

Asymptotic convergence is obtained, and from (8.26) then follows:

$$P_V^{(s)} \approx \frac{\left(k_r^{(s)}\right)^2}{V_r^{(s)\left(k_r^{(s)}+1\right)}} - \left(k_r^{(s)} - 1\right)! \tag{8.29}$$

Which in fact is close to the well-known Pareto law and is characterized by a curve of the form $V_r^{(s)}(t_k)$ from (Fig. 8.7), it indicates the presence of an "increase" ~ increment—the generation of knowledge of the sth trainee in the rth direction with an increase in training time and the presence of motivation and learning properties $B_s > 0$ and $D_s > 0$ and fading in the case of $F_r^{*(s)} \approx F_r^{(s)}$ и $F_{rs}^{(|N)} \approx F_r^{(s)}$, leading to

$\frac{dF_r^{(s)}}{dt} = \frac{dV_r^{(s)}}{dt} \to \varepsilon(0)$ and $I_{\sum}^{(s)} \to \varepsilon(0)$, especially in case of absence or attenuation of the processes of self-motivation and volitional regulation, the curve $F_r^{(s)}$ from (Fig. 8.7) coincide with the experimental data obtained in.

References

1. Burkov VN, Novikov DA (1999) Theory of active systems: state and prospects. Sinteg, Moscow, 128s
2. Kovalenko GV, Kryzhanovsky GA, Sukhoi HH, Khoroshavtsev YuE (1996) Improvement of professional training of flight and dispatching personnel. In: Kryzhanovsky GA (ed). Transport, Moscow, 320s
3. Larichev OI (2002) Theory and methods of decision-making, 2nd edn, reprint.idop. Logos, Moscow, 392s
4. Heckhausen X (2003) Motivation and activity, 2nd edn. Peter, St. Petersburg, 860s
5. Kryzhanovsky GA, Kupin VV, Plyasovskikh AP (2008) Theory of transport systems. In: Kryzhanovsky GA (ed). GA University, St. Petersburg
6. Kryzhanovsky GA, Shashkin VV (2001) Management of transport systems. Part 3. Severnazvezda, St. Petersburg, 224s
7. Kryzhanovsky GA, Chernyakov MV (1986) Optimization of aircraft transmission systems data. Transport, Moscow, p 24
8. Owen G (1971) Game theory. Mir, Moscow, 230s
9. Hermeyer YuB (1976) Games with non–contradictory interests. Nauka, Moscow, 327s
10. Borisov VV, Kruglov VV, Fedulov AS (2007) Fuzzy models and networks. Hotline – Telecom, Moscow, 284s
11. Zaitsev EN, Bogdanov EV, Shaidurov IG, Pesterev EV (2008) General course of transport: a textbook for the study of discipline and the performance of control work. SPbGUGA, St. Petersburg, p 98
12. Palagin YuI (2009) Logistics. Planning and management of material flows. In Textbook/Palagin YuI. Polytechnic, St. Petersburg, 286s
13. Galaburda VG, Persianov VA, Timoshinidr AA (1996) Unified transport system: studies for universities. In: Galaburdy VG (ed). Transport, Moscow, 295s

Conclusion

Modeling of aviation transport processes includes models of the movement of vehicles of all types of transport and their flows forming a dynamic situation, models of orientation and functioning of elements of the transport space and, what is especially highlighted in this tutorial—modeling of the processes of professional training and activity of operators and DM of transport processes, taking into account their motivation, volitional tendencies and intellectual level, while the models of vehicle movement and flows are considered already known from the courses of theoretical mechanics and applied mathematics, as the results of the application of the second form of the Lagrange equation for the movement of individual vehicles and queuing theory—when modeling flows their movements. That is why the main attention is paid to modeling the characteristics of the ergatic component of transport systems. For the first time, models of motivation dynamics of volitional tendencies and the formation of professional thinking abilities of operators and DM of active transport systems are presented. Significant importance is attached to such one of the most important types of professional training of operators and DM transport workers as simulator training. Many of the results of the above are obtained on the basis of intuitive and inductive reasoning, which is based on many years of experience in modeling transport processes. However, the development of transport and its infrastructure is rapidly gaining momentum and what is considered important today may soon be secondary. That is why we consider it useful to have a discussion component in the textbook, which is undoubtedly a fairly successful component of it.

© The Editor(s) (if applicable) and The Author(s), under exclusive license
to Springer Nature Singapore Pte Ltd. 2023

G. A. Kryzhanovsky et al., *Modeling of Transportation Aviation Processes*, Springer
Aerospace Technology, https://doi.org/10.1007/978-981-19-7607-0

Bibliography

1. Miroshnikov AN, Rumyantsev SN (1999) Modeling of control systems of technical means of transport. Elmore, GETU.–St. Petersburg, 224s
2. Venikov VA, Venikov GV (1984) Similarity theory and modeling. Higher School, Moscow, 243s
3. Miloslavskaya SV, Pluzhnikov KI (2001) Multimodal and intermodal transportation. RosKonikult, Moscow, 347s
4. Mirotin LB, Nekrasov AG (2007) Integrated management systems in transport chains. Bulletin of transport information. Inf Pract J 5(143):34–37
5. Lukomsky YuA, Peshekhonov VG, Skorokhodov DA (2002) Navigation and ship traffic control. Elmore, St. Petersburg, 360c
6. Trius EB (1967) Problems of mathematical programming of transport type. Sovetskoeradio, Moscow, 208s
7. Buslenko NP (1978) Modeling of complex systems. In: Buslenko NP (ed). Nauka, Moscow, 399s

© The Editor(s) (if applicable) and The Author(s), under exclusive license to Springer Nature Singapore Pte Ltd. 2023

G. A. Kryzhanovsky et al., *Modeling of Transportation Aviation Processes*, Springer Aerospace Technology, https://doi.org/10.1007/978-981-19-7607-0

Printed in the United States
by Baker & Taylor Publisher Services